Earth's Infinite Energy:

Bringing Order from Earth's Chaos

Alwaleed Alghanim

Earth's Infinite Energy Copyright © 2020 by Alwaleed Alghanim. All Rights Reserved.
Illustration Copyright © 2020 by jjgraphic. All Rights Reserved

All rights reserved. No part of this book may be reproduced in any form or by any electronic or mechanical means including information storage and retrieval systems, without permission in writing from the author. The only exception is by a reviewer, who may quote short excerpts in a review.

Alwaleed Alghanim
Email: alwaleedalghanim01@gmail.com

Printed in the United States of America

First Printing: Oct 2020
Amazon

ISBN-9798695412697

"Earth provides enough to satisfy every man's needs, but not every man's greed."

— MAHATMA GANDHI

"In all things of nature there is something of the marvellous."

— ARISTOTLE

CONTENTS

Contents

Preface 6
Chapter 1: History 9
   Geological History 10
      Cambrian 11
      Ordovician 13
      Silurian 15
      Devonian 17
      Carboniferous 19
      Permian 21
      Triassic 23
      Jurassic 25
      Cretaceous 27
      Quaternary 29
   Human History 31
      Antioch Earthquake 32
      Jiajing Earthquake 33
      Genroku Earthquake 34
      The 'Great Shake' 35
      The Messina Earthquake 36
      Gran terremoto de Valdivia 38
      The Great East Japan Earthquake 40
Chapter 2: The Order and Chaos Systems 42
   Composition of the Earth 43
      Chemical Composition 45
      Mechanical Composition 49
      Tectonic Plates 52
   Tectonic Plate Interactions 54
      Plate Boundaries 55
      Faults 57
      Measuring Earthquakes 59
      Tsunamis 63

- Chapter 3: Harnessing Earthquakes 64
  - Energy and Boundaries 65
    - Energy of Earthquakes 66
    - Practicality by Boundary 67
    - Mathemetical Derivation from Earthquakes 68
  - Generating Energy 76
  - Storing Energy 79
    - Definition of 'Battery' 81
    - Components of a Battery 81
    - How Does a Battery Work? 82
    - Types of Batteries 83
- Conclusion 96
- Glossary 98
- Index 111
- Bibliography 113
- About the Author 114

Preface

*"If you wish your power to grow on planet Earth, give thanks for the power you have.
Find ways of nurturing this power as you all learn to use it consciously."*

—Steve Rother and the group

LIFE TODAY SHOWS US OUR GREATEST ACHIEVEMENTS. From images of black holes and decoding brain speech signals into written text to treating HER2-positive breast cancer and genome editing, we live in a scientific and technological boom. While humanity is advancing at a rate faster than ever before, there is a downside.

While humanity is advancing at a rate faster than ever before, there is, quite unfortunately, a downside that would eventually lead to our demise. With the growing population and the ever-increasing needs of the human race, there must be a source for all what we have. A source that is quickly depleting. The downside of our lifestyle, a challenge that is increasing in awareness, is one that takes many shapes and forms. However we might look at it, and however harsh it might seem, we are destroying our planet.

This is what the book is about. Symbolizing physical sensation and growth, the earth can help in many ways that we do not yet know. Harnessing one of its deadliest forces, the earthquake, is only one of many sources of energy that would theoretically be possible in the future. In a time of growth, as shown in many ways, with scientific growth far from being an exception, we must look at the very element that represents it.

For that, the Earth could be seen to be comprised of two systems: The Order System and the Chaos System. The Order System is a system that consists of the internal processes of the Earth, such as the convection processes in the mantle. On the other hand, the Chaos System is a system that consists of the external processes, often being a result of the Order System. As such, we must look at one of the most chaotic processes in the Chaos System: earthquakes.

Chapter 1: History

Geological History

"*The history of any one part of the Earth, like the life of a solider, consists of long periods of boredom and short periods of terror.*"

—Derek Ager

HISTORY IS THE GATE TO THE TRUE HORRORS OF EARTHQUAKES. While there is much to cover in the history of our planet, it is the key to unraveling the true chaos of the Earth.

The Earth's surface is ever-changing. Representing the skin of a human, much of the planet's processes are hidden from the rest of the cosmos. While we can never observe these processes up close, the internal chaos of the Earth does reveal itself in its history. Alas, the story of the Earth's chaos is a long story. The following is the story of Earth's most chaotic ages.

Cambrian

542 to 488 million years ago

The Cambrian period, between 542 and 488 million years ago, was a period that oversaw rapid plate movement.

Before the Cambrian period, all the previous continents had briefly come together to form the supercontinent of Pannotia. By the beginning of the Cambrian, however, Pannotia had begun to break up, while another supercontinent, Gondwana, was forming. As Pannotia broke up, the Iapetus Ocean opened and widened between the continents of Laurentia (currently known as North America), Baltica (now northern Europe), and Siberia. The Panthalassic Ocean, meanwhile, lay to the north of the Laurentia. During the Early and Middle Cambrian, Laurentia swiftly drifted from the poles to the tropics near the Equator.

To the east of the other continents, Gondwana stretched from the South Pole to the Equator. Gondwana was made up of modern China, India, Australia, Antarctica, Africa, and South America. Two major land masses that were not part of Gondwana were Laurentia and Siberia. Throughout the early Cambrian, a growing mid-oceanic ridge between the two landmasses and Gondwana pushed them on a journey northward. Other smaller continental blocks existed, such as Kazakhstan and China, and part of Southeast Asia. Most small continental blocks were surrounded by shallow seas. While Laurentia drifted from polar to tropical latitudes, Gondwana rotated 90 degrees, driven by the movement of the Earth's tectonic plates. During periods of high sea levels, the continents were overwhelmed, apart from Gondwana, eastern Siberia, and Kazakhstan, all of which were mountainous.

By 500 million years ago, near the South Pole, Avalonia (which contained parts of Britain, Ireland, and the eastern American seaboard), Iberia (composed of Portugal and Spain), and Armorica (composed of other fragments of western Europe), were underwater off the coast of Gondwana, 13,000 kilometers (8,000 miles) away from their current positions.

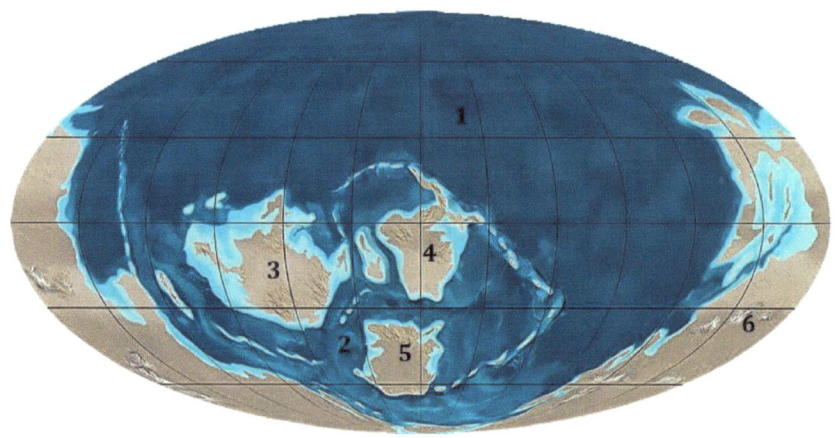

Cambrian Period
500 million years ago

1) Panthalassic Ocean
2) Iapetus Ocean
3) Laurentia
4) Siberia
5) Baltica
6) Gondwana

The Earth 500 million years ago
(Courtesy Geology Page)

Ordovician

The rapid plate movement and volcanism of the Cambrian Period continued, causing an extreme reorganization of the positions of the continents and ocean basins. Large shallow epicontinental seas surrounded the major landmasses. Rising sea levels and tectonic activity reduced the smaller landmasses to a series of archipelagos. Four major continents existed during the Ordovician, which are as following: Gondwana, Laurentia, Baltica, and Siberia.

By the Middle Ordovician, Siberia had moved from the southern to the northern hemisphere. This movement led Siberia closer to Laurentia, which lay across the Equator. The Panthalassic Ocean remained to north of both landmasses. Gondwana still extended from the north of the Equator to the South Pole. Throughout the Ordovician, Gondwana rotated anticlockwise, carrying what are now Australia and part of Antarctica into the northern hemisphere. The small islands of the South China block lay to the west of Gondwana. Baltica, in the Paleo-Tethys Ocean between Gondwana and Laurentia, moved toward lower latitudes. At the same time, the Iapetus Ocean continued to widen.

Toward the end of Ordovician period, 460 million years ago, the Iapetus Ocean started to close, while the Rheic Ocean started to open. Both seas were found on either side of thin strips of land near the South Pole that now form the eastern coastline of North America. Gondwana began to break up. The remaining parts moved to the south, so that what is now North Africa lay directly over the South Pole. The land areas of many of the continents were growing; intense volcanic activity added land to the east coast of Australia and parts of both Antarctica and South America.

The Ordovician landscape was filled with barren continents, with frequent volcanic eruptions and earthquakes, and coastlines that were constantly being reshaped. The wide, shallow seas of the Ordovician were inhabited by extensive coral reefs and a diverse number of marine invertebrates, until the end of the Ordovician when a mass extinction occurred.

Late Ordovician Period
450 million years ago

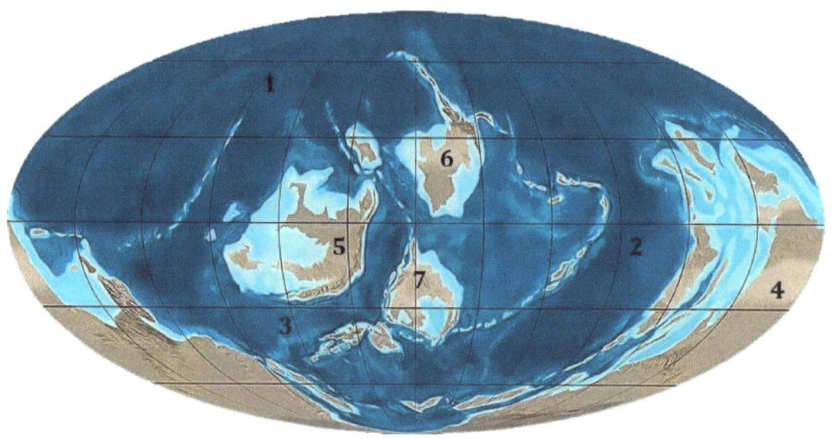

1- Panthalassic Ocean
2- Paleo-Tethys Ocean
3- Iapetus Ocean
4- Gondwana
5- Laurentia
6- Siberia
7- Baltica

> The Earth 450 million years ago
> (Courtesy Geology Page)

Silurian

420 million years ago, during the Silurian period, most of the continents lay in the southern hemisphere. The supercontinent of Gondwana, including South America, Africa, Australia, and India, was located around the South Pole. Greenland and Alaska, now near the North Pole, were on the equator.

During the Silurian period, the Iapetus Ocean began to close. Baltica and Avalonia (southern Britain and Ireland), meanwhile, moved north to collide on the southern and eastern edges of Laurentia; causing many island arcs to be displaced. Throughout the Silurian period, the Rheic Ocean opened and widened to the south of the new landmass and to the north of Gondwana.

One of the results of the northerly movement of Baltica into Laurentia was a continuous northward movement of Siberia into higher latitudes in the Panthalassic Ocean, which was shrinking. During the Silurian, the North and South China blocks started to move away from the northern edge of Gondwana, going north across the Paleo-Tethys Ocean. Gondwana rotated into an even more southerly, poleward orientation, carrying Australia to the Equator and Antarctica into the southern hemisphere. The closing of ocean basins and the rapid melting of ice sheets increased the sea levels significantly, helping to expand shallow sea environments for corals and fishes.

Silurian Period

430 million years ago

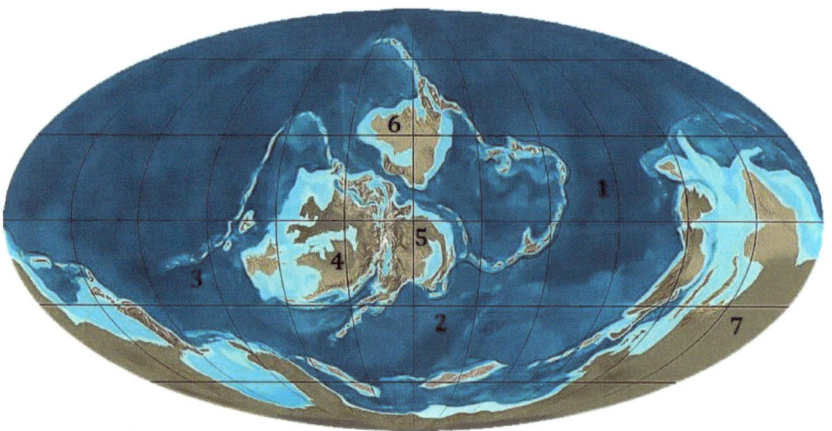

1) Paleo-Tethys Ocean
2) Rheic Ocean
3) Iapetus Ocean
4) Laurentia
5) Baltica
6) Siberia
7) Gondwana

The Earth 430 million years ago
(Courtesy Geology Page)

Devonian

By the middle of the Devonian period, 400 million years ago, Baltica and Laurentia had collided, causing the Iapetus Ocean to disappear. This collision unleashed massive tectonic forces that continued to reshape the continental landscape. This forced up the Caledonide mountains in Scandinavia, northern Britain, and Greenland; as well as the Northern Appalachian chain in eastern North America.

Gondwana rotated in a clockwise direction around an axis centered on Australia, bringing the western edge of the continent closer to the Equator and to Laurentia. Landmasses gradually became greener as ferns and tree-like plants formed forests and swamps.

360 million years ago, two supercontinents were slowly drifting toward each other. In the south is Gondwana, which consisted of Australia, Antarctica, India, Africa, and South America. In the north is Laurentia, which consisted of North America and northern Europe. Shallow seas flooded what is now the American Midwest.

Late Devonian Period

370 million years ago

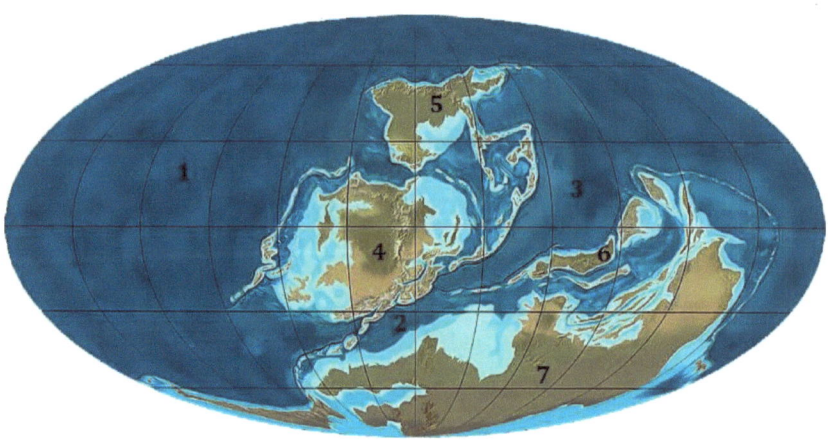

1) Panthalassic Ocean
2) Rheic Ocean
3) Paleo-Tethys Ocean
4) Euramerica (Laurentia & Baltica)
5) Siberia
6) South China
7) Gondwana

The Earth 370 million years ago
(Courtesy Geology Page)

Carboniferous

Carboniferous, which means "carbon-bearing", is a fitting name for a period when lush swamps laid down large coal deposits across the Earth. Even so, the world was largely an icehouse, with extensive ice sheets that persisted for tens of millions of years.

The Carboniferous period began 354 million years ago. Pangaea, the greatest supercontinent, was forming from the collision of Laurentia (consisting of North America and Europe) with Gondwana. Before the collision, Gondwana had rotated clockwise.

During the Early Carboniferous, Euramerica moved towards Gondwana, turning the Rheic Ocean into a narrow seaway between the western edge of Gondwana and the southwestern tip of Euramerica. The mountain-building caused by the movement peaked in the formation of the Appalachian and Variscan chains. The rotation of Gondwana opened the Tethys Ocean in the east and closed the Rheic Ocean in the west. Meanwhile, the ocean between Baltica and Siberia had started to close, paving the way for another continental collision.

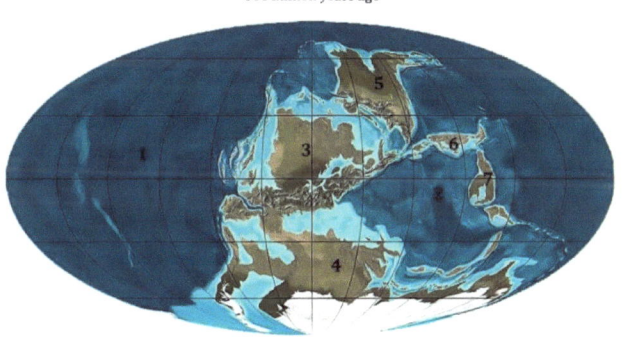

Carboniferous Period
306 million years ago

1) Panthalassic Ocean
2) Paleo-Tethys Ocean
3) Pangaea
4) Gondwana
5) Siberia and Khazakhstan
6) North China
7) South China

The Earth 306 million years ago
(Courtesy Geology Page)

The Paleo-Tethys Ocean was enclosed by Euramerica and Gondwana to its west, and the North China and South China islands to its east.

In the Early Carboniferous, a major ice cap began to develop in the South Pole. The ice caps around the South Pole spread to cover much of what was Gondwana. Meanwhile, Pangaea lay between the Panthalassic Ocean to its west and the Paleo-Tethys Ocean to its east.

Large-scale glaciation affected sea levels. As the amount of ice increases, sea levels drop. As the amount of ice decreases, sea levels drop. This has affected both the coastal and offshore habitats. There were times when seawater flooded into coastal swamps, and times when shallow bays, deltas, and inlets dried out. These fluctuations are reflected in the sedimentary deposits of the Carboniferous period.

Permian

The Permian period was the last period of the Paleozoic era, lasting from 290 to 248 million years ago. This period was dominated by Pangaea, which finally merged with Siberia, Laurentia, and Gondwana. Thereon, Pangaea stretched from pole to pole.

In low latitudes, the Central Pangaean Mountains had extended in an east-west orientation, considerably affecting atmospheric currents. Consequently, weather patterns are also affected. By the Permian period, however, the mountain belt moved further north into more arid regions. The mountains blocked the humid, equatorial winds, causing deserts to form in the northern part of Pangaea (consisting of the central part of North America and northern Europe). Most of Pangaea extended in the southern hemisphere, including the former supercontinent of Gondwana.

The end of the Permian was marked by the largest mass extinction in history. The mass extinction might have been caused by a sharp decline in sea levels that killed off most living organisms in shallow shelf areas.

In Siberia, found in the northern region of Pangaea, one of the Earth's largest volcanic eruptions covered a vast area with flood basalts, accompanied by outpourings of ash and gases, especially sulphur dioxide and water vapor.

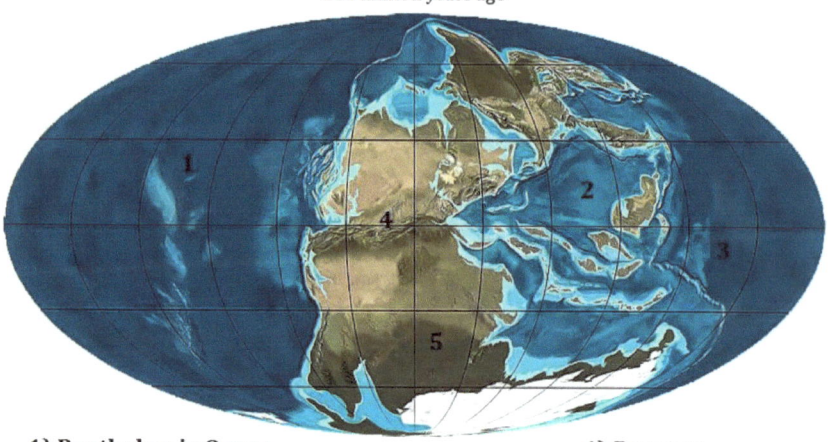

Late Permian Period
255 million years ago

1) Panthalassic Ocean
2) Paleo-Tethys Ocean
3) Tethys Ocean
4) Pangaea
5) Gondwana

The Earth 255 million years ago
(Courtesy Geology Page)

Triassic

Pangaea still made up the entirety of the world's landmasses, extending from pole to pole. Pangaea reached its peak around the Middle to Late Triassic boundary when Earth's land area was greater because of the lower sea levels. Throughout the Triassic period, Pangaea was steadily moving to the north and, as a result, Siberia moved to the North Pole. A narrow sea that separated Europe and Kazakhstania in the Permian period had closed during the Early Triassic period; causing further mountain-building in the Urals. As Pangaea moved north, it also rotated anticlockwise, taking North and South China in a northerly direction.

Meanwhile, Cimmeria moved north, crossing the Equator. This movement widened the rift between Cimmeria and the southeastern part of Pangaea, and expanded the Tethys Ocean. The Paleo-Tethys Ocean began to shrink while Cimmeria spread northward at a faster pace than the North China block, an eastern extension of Pangaea that gradually moved northward throughout the Triassic period. Pangaea (consisting of North America, Europe, North Asia, Africa, South America, India, Australia, and Antarctica) still extended near the South Pole.

During the Middle Triassic period 240 million years ago, one hemisphere was dominated by a vast ocean and the other hemisphere was dominated by Pangaea. By the end of the Triassic period, Pangaea began to break up in a process that continued for the rest of the Mesozoic era.

Early Triassic Period
237 million years ago

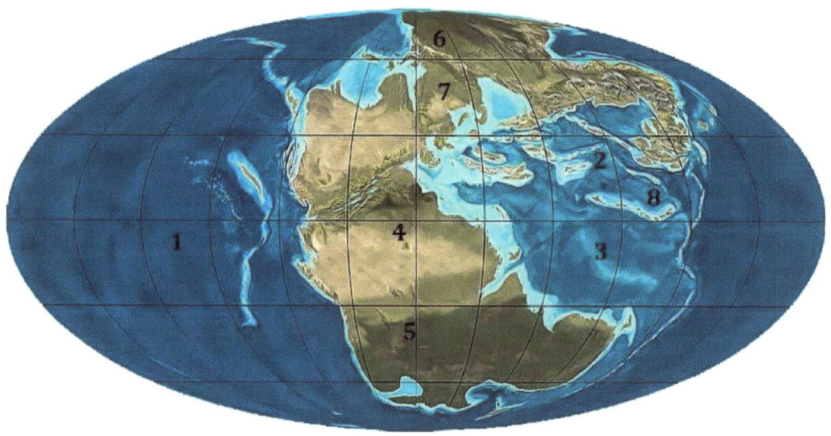

1) Panthalassic Ocean
2) Paleo-Tethys Ocean
3) Tethys Ocean
4) Pangaea
5) Gondwana
6) Siberia
7) Europe
8) Cimmeria

The Earth 237 million years ago
(Courtesy Geology Page)

Jurassic

During the Jurassic period, plate movement continued to reshape the continents and widen oceans. The separation of the northern and southern parts of Pangaea continued during the Early Jurassic period, moving in an east-west direction. This separation widened the Tethys Ocean.

In the mid-Jurassic period, the Proto-Atlantic Ocean began to open as America moved in a northwestern direction. During the later Jurassic period, Laurasia was further separated from Pangaea as North America moved away from northwestern Africa and headed north. This separation caused the western side of the Tethys Ocean to widen. 170 million years ago, both the North and South Atlantic Oceans began to open, while the Tethys Ocean began to close.

As Pangaea continued to break apart, shelf areas around the continents grew, and shallow-water environments became globally widespread. In the oceans, new types of plankton with skeletons made up of calcium carbonate and silicon appeared, forming limestone and silica-rich sediments. On the border between France and Switzerland, the Jura Mountains -after which the Jurassic is named- are formed by suvh massive limestone rocks.

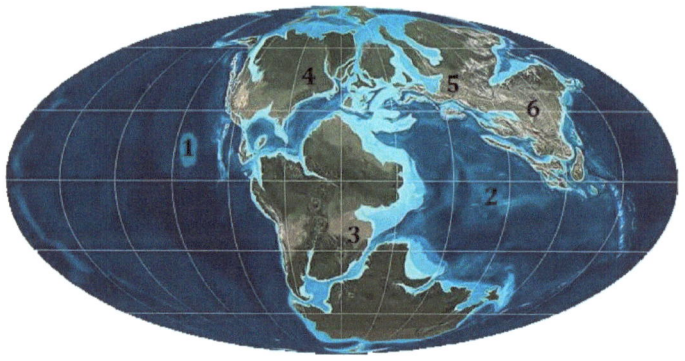

Jurassic Period
195 million years ago

1) Pacific Ocean
2) Tethys Ocean
3) Gondwana

4) Laurasia
5) Europe
6) Indochina

The Earth 195 million years ago
(**Courtesy Geology Page**)

Cretaceous

During the Cretaceous period 142 to 65 million years ago, the Earth was overall a greenhouse, with high atmospheric carbon dioxide levels and global temperatures. Sea levels reached about 200 to 300 meters (660 to 980 feet) higher than they are today. Huge, shallow seas covered much of today's continents.

The Cretaceous was a time when modern oceans started to appear more familiar. As Pangaea continued to disintegrate, Africa and South America rifted apart. As such, the South Atlantic Ocean opened. While Australia remained connected to Antarctica, India separated from the western side of Australia during the Early Cretaceous and moved in a western direction. Later, India parted from Madagascar, then rotated to move northward.

At the end of the Cretaceous, India began its collision with Asia. The collision initially caused the outpouring vast amounts of magma. This covered much of India with lava that solidified as basalt in a layer known as the Deccan Traps. Some experts suggest that the Permian and Cretaceous mass extinctions might have been caused by this type of excessive volcanic activity.

The Tethys Ocean shrank a bit as the combined blocks of Eurasia, North China, South China, and Indochina rotated clockwise. This brought southeastern Asia closer to the Equator. Sea levels were very high in the Late Cretaceous period, flooding North America and forming a seaway that extended from the Gulf of Mexico to the Arctic Ocean, which was then a newly forming ocean. The southern part of the North Atlantic, like its southern counterpart, was expanding.

Rock layers enriched with iridium, a rare element, indicate that the Earth was hit by a large asteroid by the end of the Cretaceous period. This was supported by a huge impact crater at Chicxulub, located on Mexico's Yucatan Peninsula. This event is proposed to be the cause for the mass extinction at the end of the Cretaceous.

Late Cretaceous Period
100 - 95 million years ago

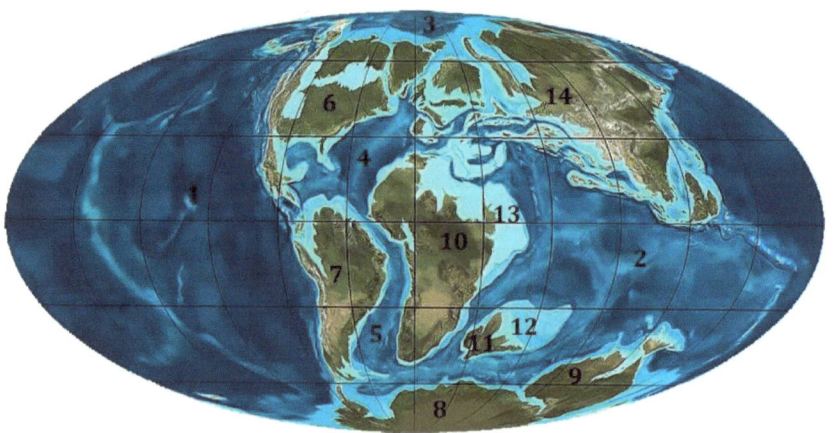

1) Pacific Ocean
2) Tethys Ocean
3) Arctic Ocean
4) North Atlantic
5) South Atlantic
6) North America
7) South America
8) Antarctica
9) Australia
10) Africa
11) Madagascar
12) India
13) Arabia
14) Eurasia

The Earth 100 million years ago
(Courtesy Geology Page)

Quaternary

The Quaternary, lasting from 1.8 million years ago to the present, has been a part of a continuing ice age. Intense periods of glaciation lasting nearly 100,000 years have been alternating with warm interglacials that lasted between 20,000 and 30,000 years.

Fluctuating global temperatures have aided in the growth and retreat of polar icecaps, continental ice sheets, and mountain glaciers; with considerable effects on the landscape. In the coldest periods, so much water was frozen that the global sea levels dropped by 100 meters (330 feet), turning shallow seas into extensions of land.

While India continued to nudge Asia, Australia into Indonesia, and both Africa and Arabia into Europe and Asia, the oceans have risen and fallen significantly. As ice sheets advanced from the north to the south, much of the water that should have returned to the oceans became locked in the frozen continents; substantially lowering the global sea levels. In Europe, the Rhine and Thames rivers joined in a large estuary (the water body in which rivers meet the seas) that emptied into the North Sea off the northern coast of England, and the English Channel did not yet exist. During the last 100,000 years of the Pleistocene, oceans levels rose, causing substantial changes to many parts of the world.

Quarternary Period
Present Day

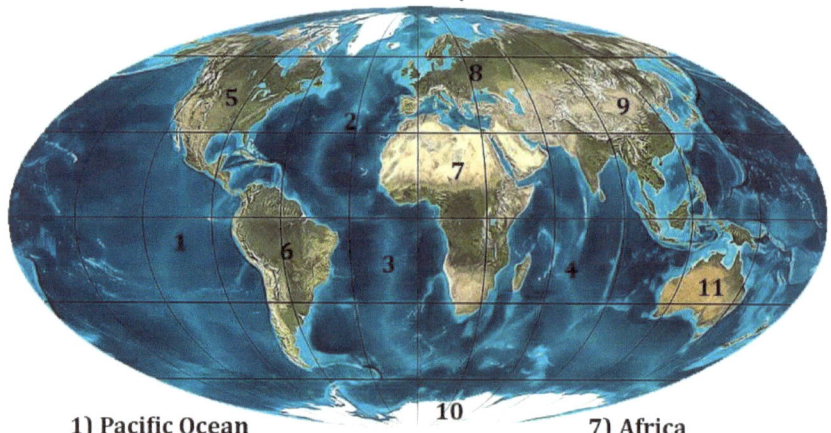

1) Pacific Ocean
2) North Atlantic Ocean
3) South Atlantic Ocean
4) Indian Ocean
5) North America
6) South America
7) Africa
8) Europe
9) Asia
10) Antarctica
11) Oceania

The Earth today (Courtesy Geology Page)

Human History

"*Ruins are reminders that while time will pass, memories remain.*"

—Sophia Khan

HUMANS HAVE ADAPTED TO MUCH OF THE CHAOS that the Earth has thrown. We have overcome an extinction event that killed hundreds of thousands, diseases that killed millions, and famine that killed tens, if not hundreds of millions of people. Yet, earthquakes always seem to surprise us, being capable of killing hundreds of thousands of people. There are many factors to these chaotic forces, yet there are still patterns. To take note of such patterns, we must take look at some of the most destructive of events.

While we may not realize it, there are hundreds of thousands of earthquakes that occur daily. There are, however, a few large earthquakes. Responsible for many other disasters, major earthquakes have killed millions, if not billions, of people. Famine and tsunamis are only two of many examples of disasters that can be caused by earthquakes. Such effects are to be considered in the following disasters

Antioch Earthquake

Fact File

Date: Late May AD 526

Location: Antakya, Turkey

Death Toll: Estimated to be the city's entire population of 250,000

Magnitude: 7.0

(Backhouse, et al. "501 Natural Disasters" 2011)

The ancient city of Antioch, found in the southeastern corner of Turkey, was an important center of early Christianity. In its most successful period, Antioch had an estimated population of half a million; only Alexandria and Rome themselves were more prestigious.

As important the city was, Antioch was in one of the most unstable areas of the planet. Antioch, located in a smaller plate in between the four plates, had a history of recurring earthquakes that damaged the buildings and fabrics of the metropolis. One such earthquake was in AD 115 when the Emperor Trajan was staying in the city before his ill-fated Parthian campaign.

Even so, the largest earthquake in Late May AD 526. With a magnitude of 7.0, this earthquake destroyed Antioch, including the great church built by Emperor Constantius II, son of Constantine the Great, 200 years beforehand. Few contemporary descriptions of the survive. Any buildings that were still standing after the initial tremor may have collapsed in a series of aftershocks. A large fire broke out the following day, consuming what was left of Antioch.

Jiajing Earthquake

The earthquake, which occurred in the winter 1556, is often referred to as the Jiajing earthquake after the Jiajing Emperor of the Ming Dynasty. The earthquake, estimated to measure at 8.0 in the Richter scale, destroyed an area of 830 sq. kilometers (520 sq. miles); in some counties, nearly two thirds of the population were killed. The epicenter of the earthquake was in the Wei River valley near Mount Hua, and aftershocks were felt for the next six months.

The main reason why this natural disaster caused such catastrophic loss of life was that a high percentage of the region's inhabitants lived in artificial caves called **yaodongs**. Yaodongs had been dug out of porous and unstable loess deposits that form the cliffs and hillsides of the area. Hundreds of thousands died when their primitive dwellings collapsed during the earthquake.

Fact File

Date: January 23, 1556

Location: Shaanxi Province, China

Death Toll: At least 830,000

Magnitude: 8.0

(Backhouse, et al. "501 Natural Disasters" 2011)

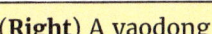

(**Right**) A yaodong

Genroku Earthquake

Fact File

Date: December 31, 1703

Location: Tokyo, Japan

Death Toll: Up to 200,000

Magnitude: 8.0

At the start of the 18th century, Edo (now known as Tokyo) was the world's largest city, with a population of a million. On the last day of 1703, however, Edo was devastated by a colossal earthquake. Most of the casualties were caused by a tsunami that smashed into a vast coastal area around Sagami Bay and the Boso Peninsula, which led the fires to sweep through the mostly wooden buildings.

(Backhouse, et al. "501 Natural Disasters" 2011)

The 'Great Shake'

The San Francisco earthquake of 1906, also known as the 'Great Shake', still provides important lessons to the study of earthquakes. In less than a minute, the earth ruptured at 13,300 kilometers per hour (8,300 miles per hour) towards the north and 10,080 kilometers per hour (6,300 miles per hour) towards the south. This cracked open 477 kilometers (296 miles) of the northern San Andreas fault from outside San Juan Bautista to the fault's triple junction found in Cape Mendocino.

One of the biggest cities in the United States collapsed. San Francisco's 400,000 residents woke up buried in the rubble of their town, and with the flames of thousands of overturned cooking stoves already catching hold. The water and gas mains had fractured, and the fire chief had been killed. San Francisco burned for three days, and the fire trebled the damage of the earthquake: 28,000 buildings were destroyed across around 500 city blocks, leaving 225,000 people homeless. In some places, groups were isolated by fire against the shore.

Fact File

Date: April 18th, 1906

Location: San Francisco, California, USA

Death Toll: 3,000+

Magnitude: 7.9

(Backhouse, et al. "501 Natural Disasters" 2011)

The Messina Earthquake

Fact File

Date: December 28th, 1908

Location: Messina, Sicily

Death Toll: 200,000+

Magnitude: 7.5

You Should Know: This earthquake is known as the deadliest earthquake in European history.

The clocks read 5:21 AM. Most people were asleep when the tremors began. For the next 30 seconds, the historic city of Messina shook violently. Meanwhile, Reggio di Calabria, an Italian town across the straits from Sicily, collapsed into rubble and dust. Buried in lumps of ragged stone and plaster in a cold, wet, and violent winter night, bewildered people had barely found their voices to scream in pain and for help when the series of tsunamis followed.

(Backhouse, et al. "501 Natural Disasters" 2011)

Three waves, rearing to a height of 6 meters (20 feet), crashed at the same time through Messina and Reggio, and thundered along the coastlines of both straits, wreaking havoc on villages and people. At Giampileri Marina, the waves peaked at 11.8 meters (39 feet). Ninety per cent of Messina was obliterated and 100,000 people immediately died.

The first shock measured 7.5 on the Richter scale. Hundreds of smaller aftershocks persisted for the following two days. With no communications or civil structure left, help came slowly. The government in Rome took five days in order to organize their lifting equipment; by which time the fleets of half a dozen Mediterranean navies were working to ferry the injured to hospitals in Naples, Rome and Malta. Their crews worked desperately, but the rubble reached 5 meters (16 feet) deep in Messina's center. The last survivors, two starving children, were dug out 18 days after the earthquake. The case was the same across the coastline.

Gran terremoto de Valdivia

In May 22ⁿᵈ, 1960, the world's most powerful earthquake, impressively titled *Gran terremoto de Valdivia*, rocked Chile. The *Gran terremoto de Valdivia* measured an unprecedented 9.5 on the moment magnitude scale (**MMS**). The epicenter of the Valdivia earthquake was located near Cañete, 900 kilometers (435 miles) south of the capital, Santiago. Even so, the earthquake was named after the city that took the biggest hit.

On May 21ˢᵗ, a day before the major earthquake, there had already been a smaller earthquake that cut off communications to Southern Chile. Rescue efforts were underway when the major earthquake struck on the following day, with terrible consequences. Two fifths of Valdivia's buildings were destroyed and many more were damaged including much of the city's industrial capacity. Both power and water supplies were knocked out, hundreds of people died, and 20,000 people were left homeless. A landslide blocked the outflow of the nearby Riñihue Lake, creating a dam that threatened to unleash a devastating flash flood. Fortunately, this danger was prevented by intense work to lower the dam and release the water in a controlled manner. Beyond Valdivia, a vast area of Chile was seriously impacted by the earthquake as coastal settlements were wiped out by tsunamis.

Fact File

Date: May 22ⁿᵈ, 1960

Location: Southern Chile

Death Toll: 2,000 – 6,000+

Magnitude: 9.5

You Should Know: This earthquake is the most powerful earthquake in known history.

(Backhouse, et al. "501 Natural Disasters" 2011)

Damage caused by the earthquake wasn't confined to Chile, however, as a tsunami raced across the Pacific Ocean with waves 10 meters (35 feet) tall recorded 10,000 kilometers (6,000 miles) from the epicenter, as far away as the Philippines and Japan. The impact was particularly severe in Hawaii, where Hilo, a coastal town, was devastated and 61 people died. Valdivia itself never actually recovered. Already subject to economic decline, Valdivia lost regional capital status to Puerto Montt in 1974. Many sites in Valdivia destroyed by the earthquake have never been redeveloped.

The Great East Japan Earthquake

Fact File

Date: March 11th, 2011

Location: Tōhuku Region, Honshu Island, Japan

Death Toll: 30,000+

Magnitude: 9.0

You Should Know: The cost of reconstruction is estimated to be $300 billion, making it the most expensive natural disaster in history.

(Backhouse, et al. "501 Natural Disasters" 2011)

In the early afternoon of March 11, 2011, the Great East Japan Earthquake exploded around 72 kilometers (45 miles) off the northeastern coast of Japan's largest island, Honshu. This undersea earthquake ranked among the world's five most powerful earthquakes in recorded history, and the worst earthquake to ever hit Japan. The earthquake lasted for six minutes and unleashed a devastating tsunami that crashed into the eastern coasts of Japan's islands, with the Tōhuku Region, located on the north of Honshu, receiving the most damage.

Waves up to 38 meters (124 feet) tall roared as far as 10 kilometers (6 miles) inland. Roads and railways were flooded, and entire towns and villages were swept away. By the time waters receded, around 125,000 buildings had been seriously damaged or destroyed. Thousands of people were either dead or missing, and millions of households in Northern Japan were left with neither electricity nor drinking water. Although Japan withstood the burden of the tsunami, the coast of Chile – 17,000 kilometers (11,000 miles) away – was still hit by a wave 2 meters (6 feet) high.

Chapter 2: The Order and Chaos Systems

Composition of the Earth

"The noble science of Geology loses glory from the extreme imperfection of the record. The crust of the earth with its embedded remains must not be looked at as a well-filled museum, but as a poor collection made at hazard and at rare intervals."

— Charles Darwin

DESPITE THE CHAOS OF EARTHQUAKES, they are organized by a system of order. Abiding by the laws of many fields of science, the system of such chaotic forces is far more complex than what can be explained by what occurs on the surface. In reality, what virtually occurs in the surface is a result of what is hidden beneath. To understand the processes that occur, we must study what lay directly under our feet: tectonic plates.

To first understand what tectonic plates are and what drives their movement, we must first look at the Earth's composition. The Earth's composition can be studied in two different ways: the chemical composition and the mechanical composition; both of which are important.

The Earth's surface, as we are aware, is made up of rocks. For some, these rocks can be seen where the layer of soil is thin. For others, these rocks can be seen in quarries and cliffs. Yet we must ask: Can the Earth be made up of the same rocks to the core?

A simple way to determine whether that is plausible is by determining the Earth's density. The rocks found on the surface have a density of 2.7 tonnes per m³. If the Earth is made up of the same rocks, then it would have a slightly greater density because of the compression at depth.

To identify the density of the Earth, we must first identify its mass and volume. The Earth's mass is known with great precision because it can be determined from the duration of the Moon's orbit. The resulting mass is 6×10^{21} tonnes (6 thousand billion billion tonnes). Since the Earth has a volume of 1.084×10^{21} cubic meters, we can deduce the Earth's density by dividing its mass by its volume. Thus, the Earth's density is 5.52 tonnes per cubic meter, more than twice the density of the Earth's surface. Hence, the Earth's interior is far denser than its surface. In fact, the Earth's core is believed to be mostly made up of iron, with a density of nearly 13 tonnes per cubic meter.

Chemical Composition

The Crust

The crust is made up of slightly less dense rock than the mantle, and has a relatively higher concentration of silicon, aluminum, calcium, sodium, and potassium; however, the crust has a relatively smaller concentration of magnesium. Essentially, the crust is just the light materials that floated to the top. There are two types of crust:

a) **Continental crust**: Mainly comprised of granite, continental crusts underlie virtually all the land surface and the shallow seas. Continental crusts vary in thickness from 25 kilometers (15.5 miles) in thin, stretched areas to as deep as 90 kilometers (55.9 miles) below the highest mountain chains, such as the Rocky and the Himalayan mountain chains. The density of continental crust is slightly less than oceanic crust at around 2,900 kilograms per cubic meter.

The variation in the thickness of the continental crust is caused by giant downward protuberances of the base of the crust. This shows that mountains are not held up by the strength of the material on which they rest. Instead, crustal areas with different thicknesses may be thought of as floating on the mantle.

b) **Oceanic crust**: Mainly comprised of basalt, the oceanic crust, forming the floor of the deep oceans, ranges in depth from 6 to 11 kilometers (3.7 to 6.8 miles) thick. The density of oceanic crust is slightly higher than the continental crust at around 3,000 kilograms per cubic meter. The oceanic crust may be denser than continental crust because of the differences in their composition (see **page 62**).

The concept of "blocks" of crust floating in a balanced state is known as **isostasy** and is applied almost everywhere. Buoyancy shows us why oceanic crust never appears on dry land. It is thinner and denser than continental crust, so it floats lower.

Studying the transfer of **seismic waves** (vibrations triggered by earthquakes) has led to several discoveries about the composition of the Earth. There are two types of seismic waves to be focused on, which are: compressional waves and shearing waves. A **compressional wave** (or **P-wave**) consists of alternate pulses of expanding and shrinking, similarly to sound waves in the air. A **shearing wave** (or **S-wave**), meanwhile, is an alternate side-to-side wobble traveling across a body; which can be observed by shaking a jelly. An important distinction between compressional and shearing waves is that while compressional waves can travel through any material, shearing waves cannot pass through liquids because they offer little to no resistance to shearing motion.

Junctions between two rock bodies are sometimes indicated by sharp changes in seismic speed. On a global scale, this effect is one of gradually increasing speed with depth because the increase in depth, and hence pressure, causes rigidity to increase. The speed of P-waves gradually increases from 7,200 kilometers per hour (4,473.9 miles per hour) just below the surface to an average of 23,400 kilometers per hour (14,540 miles per hour). Below this, there is a notably sharp jump to a speed of 28,800 kilometers per hour (17,895.5 miles per hour) at an average of 30 kilometers (18.6 miles) below the continents, but usually 10 kilometers (6.2 miles) or less below the ocean floor. The sharp change to denser rocks is known as the **Mohorovicic discontinuity**, or **Moho** for short, which is found between the crust and the mantle.

The Mantle

The compositional difference between the crust and the mantle is relatively slight, but it is sufficient to account for the sharp change in the speed of seismic waves across the Moho.

Some depth-related changes in seismic speed have been identified. However, the changes are believed to represent the pressures at which atoms within crystals become packed into denser and more rigid structures rather than changes in chemical composition.

The most abundant elements in both the crust and the mantle are silicon and oxygen. Any compound made of a chemical combination of these two elements -silicon and oxygen- is known as **silica**. Therefore, the rocky materials mainly comprised of silicon and oxygen are commonly called **silicates**.

Throughout, it is believed that the mantle has a chemical composition like that of **peridotite**, a dense, coarse-grained rock. The distinction of the composition between the core and the mantle is much more crucial than that between the mantle. Mainly comprised of iron, the core does not even consist of silicates.

Composition of the Crust and Mantle (*Percentage*)				
Element	Oxide	Continental Crust	Oceanic Crust	Mantle
Silicon	SiO_2	62	49	45
Titanium	TiO_2	0.8	1.4	0.2
Aluminum	Al_2O_3	16	16	3.3
Iron	Fe_2O_3	2.6	2.2	1.2
Iron	FeO	3.9	7.2	6.7
Magnesium	MgO	3.1	8.5	38.1
Calcium	CaO	5.7	11.1	3.1
Sodium	Na_2O	3.1	2.7	0.4
Potassium	K_2O	2.9	0.26	0.03

The Core

While the outer core can be identified as liquid because of shearing waves' inability to pass through it, the inner core can be identified by a rapid increase in the speed of P-waves. Unlike the outer core, the inner core does transmit S-waves, meaning that it is solid.

The inner core has the properties of solid iron combined with a small percentage of nickel. Despite having the properties of other metals, such as cobalt or titanium, the iron core is the most likely because that would support the fact that the Earth is rich with the same metallic elements as the Sun and meteorites. The liquid outer core, however, has a density too low to consist of pure metal. Therefore, around 10% of the outer core must be comprised of at least one relatively light element. While the exact elements cannot be proved, the most likely elements are oxygen, sulfur, carbon, hydrogen, and potassium (a light metal).

Mechanical Composition

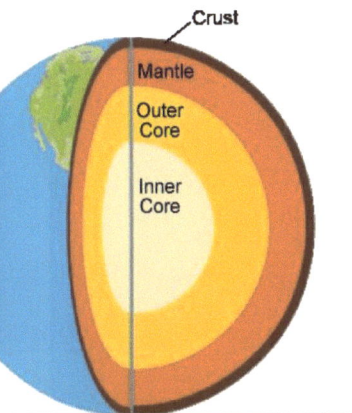

4.5 to 5 billion years ago, Earth originally had two components: the solid mass lithosphere and the surrounding gaseous atmosphere. Once the temperature of the primitive Earth cooled down below 1,000° C, the liquid components of the Earth formed what is known as the hydrosphere.

The Earth consists of four parts: the atmosphere, lithosphere, pyrosphere, and baryosphere.

The composition of the Earth (Courtesy FlinnPREP)

Atmosphere

Compared to the hefty size of the planet, the atmosphere is no more than the skin of an apple. It is so sparse as to be virtually unbreathable at the tops of the tallest mountains 8 to 9 kilometers (5 to 5.6 miles) above sea level, clouds rarely occur higher than 12 kilometers (7.5 miles), and at 200 kilometers (124.3 miles) it is so insubstantial that it offers little to no resistance to artificial satellites as they orbit.

Lithosphere

The **lithosphere** is the outermost part of the Earth, comprised of the crust and the uppermost region of the mantle. With a thickness of 32.2 to 40.2 kilometers (20 to 25 miles), the lithosphere mainly consists of silica and aluminum, being rocky in both its composition and in its mechanical properties.

Asthenosphere

Right below the lithosphere is a weak part of the mantle, despite having the same mechanical composition. Dubbed the **asthenosphere**, this layer coincides with the **low-speed layer**, a layer in which the speed of seismic waves drops significantly. Although it is solid in terms of the transmission of seismic waves, the rock below the lithosphere is not stationary, as seen in the pyrosphere.

Pyrosphere

The **pyrosphere**, also known as the mantle, is the middle part of the Earth. With a thickness of 2,896.8 kilometers (1,800 miles), the pyrosphere mainly consists of silica, manganese, and magnesium. The mantle has a plastic-like behavior, and the temperatures range from 1,000° to over 3,000° C. The difference in temperature and behavior of the mantle allows the mantle convection process to occur. **Mantle convection** is the process in which the Earth's mantle moves due to convection currents (currents in a fluid caused by the tendency of hotter materials to move upwards while colder materials sink below) carrying heat from the Earth's interior to the surface.

Baryosphere

The **baryosphere** is the central core of the Earth. It is filled with molten magma rich with iron and nickel. The baryosphere has two zones:

1) **Inner core region**: The inner core region is made up of iron and nickel, reaching temperatures of 4,300° C. The high pressures applied to this region causes it to be solid; with a 1,287.5-kilometer (800-mile) radius.

2) **Outer core region**: The outer core region is made up of liquid iron, reaching temperatures of 3,400° C. This region has a radius of 2,253 kilometers (1,400 miles).

Tectonic Plates

Tectonic plates, located in the lithosphere, are a description of how the Earth's crust is divided into continent-sized plates jostle around; creating volcanoes and mountain chains. Tectonic plates can be either **continental** or **oceanic**, both of which can be classified under three classifications: Major plates, minor plates, and microplates.

1) **Major plates** (also known as **primary plates**) comprise much of the continents and the Pacific Ocean. For simplicity's sake, any tectonic plates with an area larger than 20 million sq. kilometers would be considered as a major plate.

2) Minor plates (also known as **secondary plates**), meanwhile, are often not shown on major plate maps, as most minor plates don't cover significant land area. For simplicity's sake, **minor plates** are any plates between 1 and 20 million sq. kilometers.

3) Finally, microplates (also known as **tertiary plates**) are often grouped with adjoining major plates. For simplicity's sake, **microplates** are any plate smaller than 1 million sq. kilometers.

There are many plates on the Earth, most of which will not be discussed because of their sheer amount.

Tectonic Plate Interactions

"It takes an earthquake to remind us that we walk on the crust of an unfinished planet."

— Charles Kuralt

EARTHQUAKES SHOOK. Volcanoes roared. Tsunamis thundered. It is easy to forget that these chaotic forces are all half-siblings, given birth from the different interactions from the same tectonic forces. As such, the system of order could only unleash their chaos from such tectonic forces. To bring order from chaos, it is necessary to look at the roots of such chaotic forces.

Plate Boundaries

There are three main types of **plate boundaries** (the boundary areas between tectonic plates), which are:

Convergent boundaries

Convergent boundaries are boundaries in which two plates collide. This type of boundary can further be divided into three types:

1) Oceanic-Oceanic: This refers to the convergent boundaries comprised of two oceanic plates, where island arcs and oceanic trenches occur. Regions of active seafloor spreading can also occur behind the island arc, known as **back-arc basins**. These basins are commonly associated with submarine volcanoes.

2) Oceanic-Continental: This refers to the convergent boundaries comprised of an oceanic plate and a continental plate. The denser oceanic plate subducts (descends) under the continental crust, often forming a mountain. An example of such mountain ranges is the Andes mountain range that runs along South America's western side.

3) Continental-Continental: This refers to the convergent boundaries comprised of two continental plates. Being too light to subduct, the two plates would collide, creating especially large mountain ranges; the largest of which being the Himalayas mountain range.

Divergent boundaries

Divergent boundaries are boundaries in which two plates move apart. These boundaries can form within continents, but they will eventually open and become ocean basins. The resulting are can also be filled with crustal material from the molten magma that forms below. There are two cases to divergent boundaries:

1) **On land**: Divergent boundaries within continents initially form rifts, which then form rift valleys.

2) **Under the sea**: The most active divergent boundaries are those between oceanic plates, commonly referred to as **mid-oceanic ridges**.

Transform boundaries

Transform boundaries are boundaries in which plates slide past each other. The relative motion of the plates occurs horizontally, which means that plates are neither created nor destroyed. Because of friction, plates are not able to glide past each other. Instead, stress builds up in both plates before releasing the energy when it exceeds the threshold of the plates. Hence, earthquakes occur.

Faults

To first understand the interactions between earthquakes, we must first understand where they occur. There are many fault lines between tectonic plates, the most famous of which is the San Andreas fault in California, United States. A **fault** is a fractured weak point that marks the contact and rupture of two plates. Instead of occurring gradually, faults unfortunately tend to stick. Strain tends to build up for decades or centuries until it reaches a critical level. Once reached, everything gives at once.

As soon as a fault has given way at a point, adjoining lengths will also slip. However, movement is unlikely to occur along its entire length as faults can reach hundreds, if not thousands, of kilometers long. Excluding exceptionally large earthquakes, movement is typically restricted to a short portion of the fault. The closer to the breakpoint, the greater the energy and potential for destruction with the strongest seismic waves being generated at the initial breakpoint.

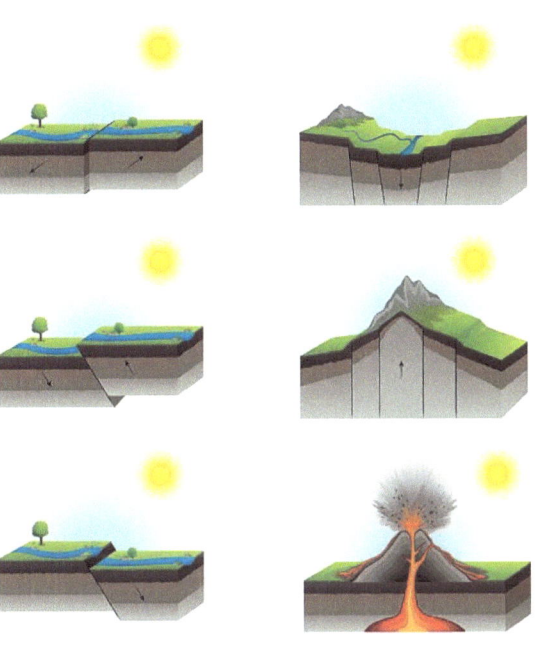

The breakpoint is called the **focus**, and the surface directly above the focus is called the **epicenter**. If the slip occurs farther along the fault, readjustments close to the focus tend to cause a series of aftershocks that continue for days (or even years in extreme cases) after the initial earthquake.

Different types of faults

Most of the damage in earthquakes is caused by **surface waves**, waves that travel along the surface of the Earth. Surface waves can be separated into two types known as Rayleigh waves and Love waves. **Rayleigh waves** are waves that roll like the waves of the ocean. **Love waves**, meanwhile, are waves that shake violently from side to side.

Measuring Earthquakes

The **Richter scale** is an attempt to determine the strength of seismic waves at the focus; comparing the signals detected at many seismometers at several distances.

The Richter scale is numerical, allowing comparative data of earthquakes to be accumulated. For each magnitude increase, the amplitude of seismic waves increases tenfold and the energy increases by around thirty-fold.

Principle of operation of a seismometer (Courtesy NASA/JPL-Caltech)

A disadvantage of the Richter scale is that it does not show how much damage earthquakes can deal. The effect that an earthquake has depends on its time and location. For example, a magnitude-8.0 earthquake in an isolated area would cause far less damage than a magnitude-6.0 earthquake whose epicenter is in a major city. A real-life example to prove this was an earthquake that occurred on May 24th, 2013. On that day, a magnitude-8.3 earthquake occurred, with a depth of 600 kilometers (372.8 miles) below the sea to the west of Kamchatka (eastern Russia) did no harm whatsoever.

The effect of an earthquake also depends on what type of material can be found near the surface. In a magnitude-7.5 earthquake in Mexico City, 1985, the soft, sandy subsoil trembled so vigorously that some buildings sank deep into it, causing many deaths. A similar earthquake that occurs below a city built on solid rock would have probably caused significantly less damage.

Magnitude range	Number per year	Example	Year	Magnitude	Deaths
8.0 +	1	San Francisco	1906	8.3	500
7.0-7.9	18	Mexico City	1985	7.9	30,000
6.0-6.9	134	Bam, Iran	2003	6.6	26,200
5.0 – 5.9	1,300	Plattsburgh, New York	2002	5.1	none
4.0 – 4.9	13,200	Dudley, England	2002	4.8	
3.0-3.9	130,000	Oakham, England	2015	3.8	
2.0 – 2.9	1,300,000	Mansfield, England	2014	2.6	
<2.0	millions	--			

An alternative to the Richter scale is the Modified Mercalli scale[1]. Ranking earthquakes according to the intensity of their effects, the Modified Mercalli scale can be used to map how concentric zones around the epicenter are affected. An alternative scale is called the European Macroseismic scale is very similar, running from 1 to 12 instead of I to XII.

Some of the descriptions on the Modified Mercalli scale are applicable with no exceptions, such as whether an earthquake can be felt by the people walking or by those laying down. However, the amount of damage buildings receive depends on how they were constructed. Thus, most earthquake-prone regions have building codes designed to ensure that structures are relatively collapse-proof.

The **moment magnitude scale** (abbreviated as **MMS**) is a type of magnitude scale used by seismologists to measure the size of earthquakes compared to the energy they release. Introduced in 1979 by Tom Hanks and Hiroo Kanamori, the moment magnitude scale is the successor to the Richter scale.

---

[1] See page 60

The MMS is based on the **seismic moment** of an earthquake, equal to rigidity multiplied by the average amount of slip on the fault and area. The average seismic moment of earthquakes per magnitude is shown in **Fig 2.3** (see **page 79**).

Mercalli intensity	Description
XII	Complete damage, with objects thrown in the air.
XI	Few, if any, masonry structures remain standing, bridges are destroyed, and rails are considerably bent.
X	Some well-built wooden structures are destroyed, most masonry and frame structures destroyed with foundations, and rails are bent.
IX	The damage is considerable in specially designed structures, well-designed frame structures thrown out of plumb, great damage in substantial buildings with partial collapse, and buildings are shifted off foundations.
VIII	There is slight damage in specially designed structures, considerable damage in ordinary substantial buildings, with partial collapse, and great damage in poorly built structures.
VII	The damage is negligible in buildings of good design and construction, yet it is significant in poorly built structures.
VI	Felt by everyone, many are frightened, and some heavy furniture is moved.
V	Felt by everyone, and many are awakened, and unstable objects are overturned.
IV	Felt by many people indoors and a few outdoors.
III	Felt notably by people indoors, especially on upper floors.
II	Felt only by a few people at rest.
I	Generally detected by instrument only.

Magnitude	Moment (M_0)
1	3.548133×10^{17}
2	1.122018×10^{19}
3	3.548133×10^{20}
4	1.122018×10^{22}
5	3.548133×10^{23}
6	1.122018×10^{25}
7	3.548133×10^{26}
8	1.122018×10^{28}
9	3.548133×10^{29}

Tsunamis

If the focus is found near the seafloor, the sudden displacement of rock can create a series of waves that travel at speeds of several hundred kilometers per hour. In deep water, meanwhile, such waves may barely be noticeable as they reach less than a meter high. Popularly (but incorrectly) known as tidal waves, this phenomenon is more correctly referred to as tsunami; which can also be triggered by landslides or undersea volcanic eruptions. In summary, a **tsunami** is a large wave caused by disturbances in the sea.

Chapter 3:
Harnessing Earthquakes

Energy and Boundaries

"If you want to find the secrets of the universe, think in terms of energy, frequency, and vibration."

—Nikola Tesla

THE POTENTIAL OF ENERGY VARIES BETWEEN EACH SOURCE. However, that potential is limited by many factors, such as the efficiency of the device and the abundancy of the source. As such, there are many different limitations when harnessing energy, especially renewable energy.

When considering the Chaos System, there are many different limitations because of its nature. However, as we will see, there are several ways in which most of these limitations can theoretically and mathematically be subdued.

Energy of Earthquakes

When it comes to earthquakes, there are several factors that come into play. Factors include the size of the fault, its depth, and its magnitude. Despite this, the **moment magnitude scale** (the scale that measures the relative size of earthquakes) shows the typical amount of energy per magnitude by using the formula:

$$10^{1.5m}$$

In this formula, m represents the magnitude where an increase in magnitude in the moment magnitude scale's logarithmic scale corresponds to a $10^{1.5}$ (around 32) times increase in the amount of energy released. An increase of two magnitudes corresponds to a 10^3 (1,000) times increase in energy. For example, an earthquake with a magnitude of 5.0 has 32 times more energy than an earthquake with a magnitude of 4.0 and 1,000 times that of a magnitude-3.0 earthquake.

Magnitude	Energy (Joules)
1	794,328
2	25,118,864
3	794,328,235
4	25,118,864,315
5	794,328,234,724
6	25,118,864,315,096
7	794,328,234,724,288
8	25,118,864,315,095,801
9	794,328,234,724,281,502

Practicality by Boundary

Convergent Boundary

Convergent boundaries, being responsible for most earthquakes, may seem like the perfect type of boundary to use for renewable energy. However, there is a crucial flaw: subduction.

Every year, tectonic plates tend to move at rates of 2.3 to 11 centimeters (~1 to 3 inches) per year. Therefore, if we were to use a device to harness the energy for half a decade, it would descend from 11.5 to 55 centimeters (4.5 to 21.6 inches) from where it was placed; meaning that convergent boundaries are impractical.

Divergent Boundary

Responsible for a few, if any, earthquakes, divergent boundaries tend to separate at the same rate as convergent boundaries. Thus, it is just as impractical as convergent boundaries.

Transform Boundary

Transform boundaries, the last type of boundaries, are perhaps the only practical type of boundaries to use earthquakes as renewable energy. Because plates slide past each other in transform boundaries, the crustal plate is neither generated nor destroyed. Thus, the fault lines are more fixated than those of convergent and divergent boundaries.

Instead of continuously moving, force is built up for decades, or even centuries, in transform boundaries before it gets released. Thus, both the stress and strain of earthquakes in a specific transform boundary must be calculated.

Mathemetical Derivation from Earthquakes

The San Andreas Fault, the most well-known fault line in California, is perhaps one of the most important fault lines in the study of seismology. Perhaps the most infamous earthquake in the San Andreas fault is the '*Great Shake*' (see **page 46**). This earthquake gives great insight to the static friction force of the San Andreas fault.

The '*Great Shake*', with a magnitude of 7.9, cracked open 477 kilometers (296.4 miles) of the northern San Andreas fault. The depth of the crack, meanwhile, is around 8 kilometers (5 miles. According to (Sanandreasfault.org, n.d.), fault lines vary in width from less than a mile to several miles long. Assuming that the width of the northern San Andreas fault is 1.6 kilometers (1 mile), then the volume displaced is roughly 6.14 billion cubic meters. With this data, it is possible to determine both the stress and strain of the fault.

Stress, in terms of physics, is the amount of force (in Newtons) per unit of area (in sq. m), with the unit of Newtons/m². With the symbol σ, stress is equal to the amount of force divided by the cross-sectional area. Because the majority of the energy moves in a linear movement and the fault is rectangular in shape, it is safe to assume that the most efficient and likely area is that of a rectangle.

$$\sigma = \frac{F}{A}$$

The energy of the '*Great Shake*' can be calculated using the moment magnitude formula; where E is energy and M is the magnitude of the earthquake.

$$\log E = 5.24 + 1.44M$$

A joule, in itself, cannot be converted to newtons. By definition, a unit of force can only be converted into a unit of energy by applying it in a distance. Thus, what can be converted to newton is joules/meter. To do this, we must divide the moment magnitude equation by the distance, or length in which it occurs. The moment magnitude formula can, therefore, be rewritten as the following:

$$\frac{\log E}{l} = \frac{(5.24 + 1.44M)}{l}$$

In order to get rid of the logarithmic scale, we must raise both sides to the power of 10.

$$\frac{E}{10^l} = 10^{[\frac{(5.24+1.44M)}{l}]}$$

By multiplying both sides by 10^l, we receive the stress equation in the form of energy.

$$E = 10^{l+[\frac{(5.24+1.44M)}{l}]}$$

In order to get the force, we must once again divide the equation by the distance. Because the subsequent result is $\frac{E}{l}$, it is equal to the force. Thus, by subistituting $\frac{E}{l}$ for F, the force of the 'Great Shake' can finally be calculated (**Equation 3.1**).

$$F = \frac{10^{l+[\frac{(5.24+1.44M)}{l}]}}{l} \quad \text{(Equation 3.1)}$$

Because most of the energy moves in a linear movement and the fault is rectangular in shape, it is safe to assume that the most efficient and likely area is that of a rectangle.

$$\sigma = \frac{10^{l+[\frac{(5.24+1.44M)}{l}]}}{\frac{l}{l \times w}} \quad \text{(Equation 3.2a)}$$

Strain, meanwhile, is the extension of a material's length, where Δl is the change in length and l_o is the original length of the material.

$$\varepsilon = \frac{\Delta l}{l_o} \quad \text{(Equation 3.3a)}$$

The change in length can be calculated by using the moment equation, which describes the length of rock displaced. Thus, by means of substitution, the following equation is derived where R is rigidity of the rock, A is the area, and M_o is the magnitude.

$$\varepsilon = \frac{[\frac{(R \times A)}{M_o}]}{l_o} \quad \text{(Equation 3.3b)}$$

According to Uri S. ten Brink's journal article, "Volcano spacing and plate rigidity", rigidity can be calculated using the thickness (T) of the crustal plate.

$$R = T^{\frac{3}{4}}$$

By substituting A for the area of a rectangle, and multiplying the equation by the reciprocal of the denominator ($^1/_{l_o}$), the simplest form of the strain equation can be derived (see **Equation 3.3c**).

$$\varepsilon = \frac{T^{\frac{3}{4}} \times w}{M_o} \quad \text{(Equation 3.3c)}$$

When combining the two equations, the young's modular of the *'Great Shake'* can be calculated. Young's modular represents the stiffness of the material, which in our case, are tectonic plates. The young's modular equation is equal to:

$$E = \varepsilon \times \sigma \quad \text{(Equation 3.4a)}$$

Using Equations **3.2** and **3.3**, deriving the young's modular equation requires substitution. By substituting the stress and the strain for their respective equations, the young's modulus is expanded significantly.

$$E = \frac{10^{l+[\frac{(5.24+1.44M)}{l}]}}{l^2 \times w} \times \frac{T^{\frac{3}{4}} \times w}{M_o}$$

(Equation 3.4b)

By means of simplification, dividing the width in the strain equation by the width in the stress equation would remove it from the equation. Thus, by multiplying both sides, we can derive the simplest form of the young's modulus equation (see Equation 3.4c).

$$E = \frac{10^{l+[\frac{(5.24+1.44M)}{l}]} \times T^{\frac{3}{4}}}{l^2 \times M_o}$$ (Equation 3.4c)

Using the 'Great Shake' to represent the significant earthquakes in the San Andreas fault, then the magnitude (M) equals 7.9 and the rupture length equals 9.7 meters. The thickness of the Pacific Plate is at least 7 kilometers. Thus, by substituting for the young's modulus equation, the surface has a young modular of around 515.72 newtons per square meter (0.075 psi) with an error rate of ±47.5%. With this low young's modular, how do the crustal plates handle the forces of earthquakes such as the 'Great Shake'?

The answer to this question may be amplitude. Because earthquakes occur in the form of seismic waves, the **amplitude** (the maximum difference of an alternating potential from the average value) of earthquakes can easily be calculated. The amplitude of seismic waves follows what is known as the inverse-square law. **Inverse-square law** is a physical law that states that intensity is inversely (oppositely) proportional to the square of the distance from the source.

Using the inverse-square law, the surface may face up to $\frac{1}{49,000,000}^{th}$ of the force in the focus of many earthquakes in the San Andreas fault. Thus, the Young's modulus of the Pacific Plate is 25,270,304,507.4 newtons/m² (3,665,147.8 psi), and, by comparing it to the the most powerful earthquake (the 1960 Valdivia earthquake), can reach upwards to 37,182,404,184.4 newtons/m² (5,392,851.8 psi). This is slightly higher than the typical Young's modulus of basaltic rocks than tend to reach up to 19.8067×10^7 newtons/m². The difference can be boiled down to heat and pressure. Because of the massive size of the Pacific plate, it has a lot of pressure towards the lower half, only encouraged by the Pacific Ocean above it; thus increasing the Young's modulus. This inflation in the Young's modulus is only decreased from the heat of the mantle below it, reaching up to 900° Celsius (1,652° Fahrenheit).

The pressure and heat of the Pacific plate greatly impact the Young's modulus because, on the smallest scale, they impact the atomic structure of the rocks. When pressure is increased, the atoms get compressed and form stronger bonds as they cluster together. On the contrary, when temperature is increased, the atoms are encouraged to move more freely and with more energy. This can be best exploited by boiling water, as the increase of temperature causes water vapor.

What can be inferred from the Young's modulus is that pressure is a far more important factor than heat when going with depth. This can be seen with scuba divers. In the ocean, divers face increased pressures and lower temperatures as they dive deeper. However, as important as temperature is, pressure is a far more urgent factor in many cases.

Pressure also plays a huge role during earthquakes. In order to take advantage of the Chaos System -that is to say, the unpredictable nature of earthquakes-, we must be capable of handling the pressures of the most dangerous earthquakes.

In order to know the full extent of the pressure, we must start with the basic formula for pressure, where P is pressure, F is force, and A is area.

$$P = \frac{F}{A}$$

The force, unlike in the stress equation (see page 68), cannot be directly substituted for the moment equation. Because we must utilize transform faults, there are two main forces that impact pressure: the force exerted in each plate, and the friction in between the two. Thus, we must find the net force by utilizing friction. Friction (F_k) can be calculated by using the net force (F_{net}) and the coefficient of friction (μ_k).

$$F_k = \mu_k \times F_{net}$$

Because F_{net} in this case is the total force in the tectonic plates, it can be substituted for the moment equation. The coefficient of friction in the crust/mantle interface is equal to 2.7. Thus, the force can be calculated (see Equation 3.5)

$$F_k = 2.7 \times 10^{[\frac{(5.24+1.44M)}{l}]} \quad \text{(Equation 3.5)}$$

Finally, using the area of a square as a base for what is akin to an "earthquake panel", the pressures applied in the panel can be easily calculated (see Equation 3.6).

$$P = \frac{2.7 \times 10^{[\frac{(5.24+1.44M)}{l}]}}{l^2} \quad \text{(Equation 3.6)}$$

This equation, which is quite important in the next sections, makes it way through this section to showcase one important point about the pressures and characteristics of earthquakes in the San Andreas fault. By using the puncture length of the 1906 San Francisco earthquake (9.7 m), an increase in magnitude would lead to an increase the force by 1.4 Newtons. While seemingly insignificant, this pressure proves that every part of the Chaos System is interconnected, and this raises huge challenges to harnessing Chaos System. While there is an order in ths system, we require panels consisting of materials that are able to handle changes in both pressure and heat. Not only that, but the amount of materials required will be exceptionally large. Thus, we must look into how it might be possible to harness the energy of earthquakes.

Generating Energy

"The future is green energy, sustainability, renewable energy."

—Arnold Schwarzenegger

THE ENERGY OF EARTHQUAKES, as we have seen, is one of the powerful forces to come out of the Chaos System. While our current technological and scientific advancements do not allow order to be brought to the earthquake processes of the Chaos System, there is one way that could work mathematically.

One way to harness the energy of earthquakes is to use what are known as triboelectric generators. **Triboelectric generators** are generators that use the triboelectric effect. The triboelectric effect is an effect where certain materials become electrically charged after coming into a static contact with another material and then separated.

In triboelectric generators, two sheets of **polyester** (a type of large molecules composed of many repeated subunits that contain the ester functional group in their main chain) and **polydimethylsiloxane** (simplest member of the group of silicone polymers) act as **electrodes** (conductors through which electricity enters or leaves an object) as they are rubbed together. After rubbing the two sheets, they are instantly separated. This creates a gap between them that separates the electrons on the polydimethylsiloxane sheet and generate both positive and negative charges. The two electrodes should be made of different materials in order to create a charge. If an electrical charge is connected between two electrodes (sheets), there will be a small flow of charges to equalize the charge. Thus, a small amount of alternating current can be produced. An external source (a small fraction of the Pacific and North American plates in this case) is used to press and slide the two sheets together.

Despite being intended for much smaller purposes, triboelectric generators do indeed have some potential. Georgia Tech engineers state that their triboelectric generators get as much as 300 W from a single layer of polymer at an area measuring one sq. meter; meaning that the volume power density reaches more than 400 kilowatts/m³ and efficiencies above 50%.

The 'Great Shake', at its focus, has 707,945,784,384.1373 joules of energy. Using the inverse-square law, the energy that reaches the triboelectric generators is equal to 3,146,425.7084 joules. Some of the problems that could occur from such high energy include high pressures that could damage the generators and the capabilities of a single layer in such circumstances.

However, to truly know the extent of the energy we must utilize, we must know how much energy earthquakes generate per year. Using the Young's modulus equation (see page 71), the energy of earthquakes per magnitude (see page 66), and the number of earthquakes that occur per year, the energy of earthquakes comes up to be 663.2324442 petajoules ($6.632324442 \times 10^{17}$ joules), powering New Zealand for 275 days; replacing 15.85 million tonnes of crude oil, which is equivalent to 26 Prelude FNLGs, the largest ship ever constructed, or 2.5 million elephants.

The total surface area of the 'panels' needed can be determined from the pressure equation (see page 75). By isolating the area, the pressure equation can be rewritten.

$$l^2 = \frac{2.7 \times 10^{[\frac{(5.24+1.44M)}{l}]}}{P}$$

By applying the magnitude of the 1960 Valdivia Earthquake (9.6) and a puncture length of 9.7 meters, the numerator can be calculated.

$$l^2 = \frac{249.30133}{P}$$

By using the energy of the 'Great Shake' and the area impacted by the earthquake, the average amount of pressure is 144 N/m². Thus, the area of the 'panel' is at least 1.73126 m² to be able to sustain the pressure.

As a single layer measuring one sq. meter can generate 300 watts, then the total power is equal to 519.38 watts. This power, while small, shows that it is possible to retrieve high amount of powers without pressure being an issue. While the size of the panels will vary depending on their location and the magnitudes that they face, a panel with a surface area of one sq. mile (roughly 2.59 million sq. meters) would generate nearly 777 megawatts of power (6.475 megawatt-hours) – enough to power Sweden for over 4 days. Thus, it would take 90 panels to power the entire country of Sweden for a year. While the initial cost is high, reaching up to 113,924,477.02 US dollars per kWh, the operation costs are little to none. Thus, on the long run, the low operation costs and high amount of energy would save money, space, and drastically lower carbon emissions. However, the high amount of energy raises the question: how can it be stored?

Storing Energy

"We are storing up potential problems for future generations. We really need to pursue renewable energy sources and efficiencies in energy usage to the maximum."

— Paul Johnston

WITH THE BULK OF CHALLENGES THAT ARISE WHEN HARNESSING THE CHAOS SYSTEM, could storing energy be one of them? While most areas surrounding such an impossible feat are not well-documented, especially when dealing with tectonic plates, this is perhaps one of the most well-documented areas. Despite the well documentation of storing energy, it requires, like all other areas, usage of current technology in new ways.

One way to store the energy of earthquakes is the use of large batteries. Currently, the largest virtual battery is one installed by the department of energy in Abu Dhabi in January of 2019. This battery plant is a 108 MW/648 MWh sodium-sulfur battery plant. For comparison, its storage capacity is about five times that of the lithium-ion battery system (the previous record holder) Tesla had installed in Hornsdale, Australia in 2017.

The battery plant is a set of ten batteries in different locations in Abu Dhabi. However, the batteries can all be controlled as a single plant. Hence, the department of energy calls it a "virtual" battery plant.

Instead of using lithium-ion batteries like Tesla, the United Arab Emirates used sodium-sulfur batteries from Japan's NGK. One reason for this might be that lithium-ion batteries require air-conditioning to main the right operating temperature. Sodium-sulfur batteries, on the other hand, operate at 300°C (572°F) and are heavily insulated.

The size of this virtual battery plant is so large that it could provide up to six hours of backup power to Abu Dhabi in case the electricity grid goes down. Thus, an NGK representative states a benefit of sodium-sulfur batteries in such circumstances: Under such long-duration circumstances, the sodium-sulfur batteries become cheaper than lithium-ion batteries. To truly understand whether this type of battery, if we shall use batteries at all, we must understand how batteries work.

Definition of 'Battery'

According to Antoine Allanore, a postdoctoral associate at MIT's Department of Materials Science and Engineering, "*A battery is a device that is able to store electrical energy in the form of chemical energy, and convert that energy into electricity*".

A **battery**, in its simplest form, is a self-contained device that can convert a limited amount of chemical energy into electricity.

Components of a Battery

There are three main components in a battery: There are two terminals made up of different chemicals (which are typically metals) called the anode (–) and the cathode (+). The **anode** is the terminal in which **electric current** (the rate in which electric charge flows past a point) flows in from outside. **Cathode**, meanwhile, is the terminal in which electric current flows out. What separates the two terminals is known as the electrolyte. The **electrolyte** is a chemical channel that allows the flow of electric charge from the cathode to the anode.

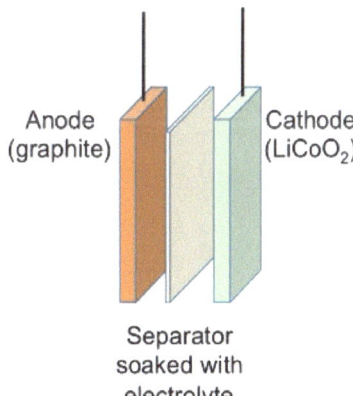

(**Right**) A simple diagram of the main components of a battery (**Courtesy** Center for Sustainable Nanotechnology)

How Does a Battery Work?

During **electric discharge** (the release and transmission of electricity through a channel), the chemical on the anode releases electrons to the terminal and the ions of the electrolyte and through what is called an oxidation reaction. An **oxidation reaction** is a reaction in which oxygen is gained. Slowly, the chemicals inside are converted into other substances. **Ions** (atoms with an electric charge due to gaining or losing electrons) are formed from what was found in the anode and cathode and take part in chemical reactions. Meanwhile, the cathode -the positive terminal-, accepts electrons, completing the circuit. The electrolyte is used to put the chemicals of the anode and cathode into contact in a manner that the **chemical potential** (the ability of an uncharged atom or molecule to work) can balance from one terminal to the other. Thus, the stored chemical energy turns into useful electrical energy.

If the battery is disposable, it will produce electricity until it reaches the same chemical potential on both terminals. This type of battery only works in a single direction, turning chemical energy to electrical energy. In other types of batteries, however, the reaction can be reversed. For example, rechargeable batteries — such as the batteries that we use in cell phones and cars — are designed in a manner in which electrical energy from an external source can be applied to the chemical system and reverse its operation; thus, restoring the battery's charge. Examples of such external sources are phone chargers and the dynamo in a car.

Types of Batteries

Batteries can generally be classified into different categories based on factors ranging from chemical composition, size, and use cases. Overall, however, there are two major types of battery:

1- Primary batteries
2- Secondary batteries

Primary Batteries

A **primary battery** is a battery that **cannot be recharged** once depleted, consisting of electrochemical cells whose reaction cannot be reversed. Primary batteries exist in several different forms, ranging from coin cells to AA batteries. They are commonly used in applications where charging is impractical or impossible, such as the batteries in military-grade devices and battery-powered equipment. It will be impractical to use rechargeable batteries in this case as recharging a battery will be the last thing in the mind of soldiers. Primary batteries always have high specific energy and the systems in which they are designed to consume a low amount of power to enable the battery to last for as long as possible. Some other examples of devices using primary batteries include pacemakers, animal trackers, wristwatches, remote controls, and children's toys.

The most popular type of primary batteries is **alkaline batteries**. Alkaline batteries have a high specific energy and are environmentally friendly, cost-effective and do not leak even when fully discharged. They can be stored for several years, have a good safety record. The only disadvantage to alkaline batteries is their low load current; limiting the use of alkaline batteries to devices that require low currents such as remote controls and flashlights. However, the application of primary batteries and their inability to recharge makes them impossible to work with.

Secondary Batteries

Also called a rechargeable battery, a **secondary battery** is a battery with electrochemical cells whose chemical reactions can be reversed by applying a certain voltage to the battery in the opposite direction. Secondary cells, unlike primary cells, can be recharged after the energy in the battery has been used up.

Secondary batteries are usually used in high drain applications and other circumstances where it will be either too expensive or impractical to use single charge batteries, as in the case of mobile phones and other gadgets and appliances. Heavy-duty batteries, meanwhile, are used in powering diverse electric vehicles and other high drain applications, such as load leveling in electricity generation. Secondary batteries are also used as standalone power sources alongside inverters to supply electricity. Although the initial cost is always more than that of primary batteries, secondary batteries are the most cost-effective over the long-term.

Secondary batteries can be further classified into several types based on their chemistry. The chemistry of secondary batteries is very important because it determines some of the attributes of the battery. A few of these attributes include the battery's specific energy, its cycle life, its shelf life, and its price. Of the many several types of secondary batteries, there are five major ones:

1. Nickel cadmium batteries
2. Nickel-metal hydride batteries
3. Sodium sulfur batteries
4. Lithium ion batteries
5. Lead acid batteries

Nickel cadmium Batteries

A **nickel cadmium battery** is a type of secondary battery developed using nickel oxide hydroxide and metallic cadmium as electrodes. Nickel cadmium batteries excel at maintaining their voltage and holding their charge when they're not in use. However, the downside of nickel cadmium batteries is their vulnerability to the dreadful "memory" effect when a partially charged battery is recharged, lowering the future capacity of the battery.

Compared to other batteries, nickel cadmium batteries offer a good life cycle and performance at low temperatures with a fair capacity. The most significant advantage of nickel-cadmium batteries is their ability to deliver their full rated capacity at high discharge rates. Nickel cadmium batteries range in sizes from AAA to D, and are used either individually or in packs of two or more cells. The small packs are used in electronics and toys, whereas the bigger packs are applied in aircraft starting batteries, electric vehicles, and standby power supply.

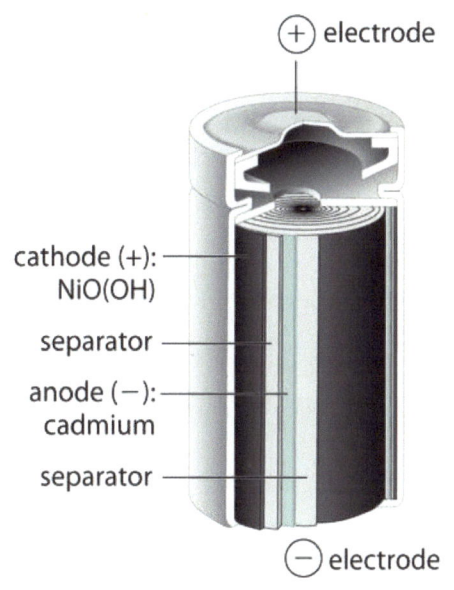

A simple diagram of a nickel cadmium battery (Courtesy Chemistry LibreTexts)

Properties of nickel cadmium batteries:

- **Energy Density**: 50 - 150 W-h/L
- **Specific Power**: 150 W/kg
- **Operating Temperature**: 0 - 30°C
- **Cycle durability/life**: 2000 cycles
- **Cost**: $7.5 per kWh

Nickel-Metal Hydride Batteries

Another type of chemical configuration used for secondary batteries is nickel metal hydride. The chemical reaction at the positive electrode is similar to that of a nickel cadmium cell, with both types of battery using nickel oxide hydroxide. However, the negative electrodes in the nickel-metal hydride battery use a hydrogen-absorbing alloy instead of cadmium.

Nickel-metal hydride batteries are applied in high drain devices because of their high capacity and energy density. A nickel-metal hydride battery can contain two to three times the capacity of a nickel-cadmium battery of the same size, and its energy density can reach to that of a lithium-ion battery. Unlike the chemistry of a nickel-cadmium battery, batteries based on the nickel-metal hydride chemistry are not susceptible to the "memory" effect.

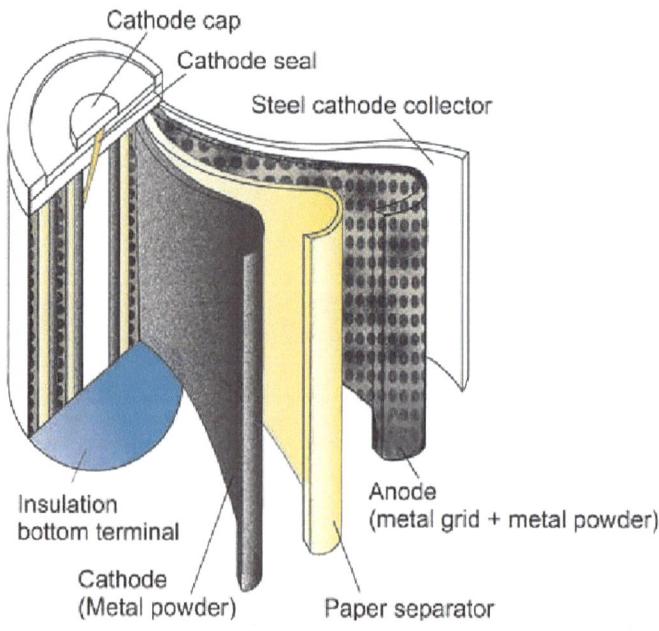

A simple diagram of a nickel-metal hydride battery (Courtesy ResearchGate)

Properties of the nickel-metal hydride batteries:

- **Energy Density**: 140 - 300 Wh/L
- **Specific Power**: 250 - 1000 W/kg
- **Operating Temperature**: 20°C
- **Cycle durability/life**: 180 - 2000 cycles
- **Cost**: $300-600 per kWh

Sodium Sulfur Batteries

The **sodium sulfur battery** is a high-temperature battery, operating at 300°C (572°F). There are two electrodes in the sodium sulfur battery: molten sulfur (+) and molten sodium (−). It is the reaction between these two electrodes that is the basis for the cell reaction. The two electrodes are separated by a solid sodium alumina, which serves as the electrolyte; allowing the positively charged sodium-ions to pass through.

During discharge, electrons are stripped off the sodium metal — an electron for every sodium atom. This leads to the formation of sodium ions that move through the electrolyte to the molten sulfur electrode (the cathode). The electrons that were stripped off the sodium metal move through the circuit and then back into the battery at the cathode. There, the electrons are taken up by the molten sulfur to form polysulfide. The positively charged sodium ions moving into the cathode balance the electron charge flow. When charging, this process is reversed. The battery must be kept at high temperatures of over 300°C (572°F) to facilitate the process. Independent heaters are thus part of the battery system. Generally, sodium sulfur batteries are highly efficient at around 89% efficiency.

Modern sodium sulfur technology was developed in Japan by the Tokyo Electric Power Co. in collaboration with NGK insulators. It is these two companies that have

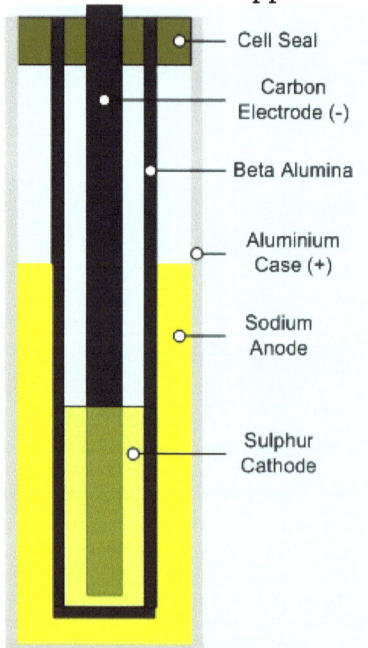

A simple diagram of a sodium sulfur battery (**Courtesy Solavox**)

commercialized such technology. Typical units have a rated power output of 50 kW and 400 kWh. Sodium sulfur batteries have a long lifetime of 15 years, or 4,500 cycles. The last feature that makes sodium sulfur batteries unique is that they are considered to have one of the fastest response times, with a start-up speed of 1 millisecond. The properties of sodium sulfur batteries are as follows:

- **Energy Density**: 150 - 240 W-h/L
- **Specific Power**: 90 - 230 W/kg
- **Temperature**: 350°C
- **Cycle durability/life**: 1500 - 4500 cycles
- **Cost**: $300 per kWh

Lithium Ion Batteries

Lithium ion batteries, one of the most popular types of secondary batteries, are a type of secondary battery in which lithium ions from the negative electrode migrate to the positive electrode during discharge and migrate back to the negative electrode when the battery is being charged. Lithium ion batteries use an intercalated lithium compound as a single electrode material. They are found in several portable appliances, including mobile phones and other smart devices. Lithium ion batteries are also applied in aerospace and military applications because of their lightweight nature.

Lithium ion batteries generally contain high energy density, a low self-discharge, and is the least vulnerable to the "memory" effect. Their chemistry, alongside performance and cost, vary according to their uses. For example, the lithium ion batteries used in handheld electronic devices are generally based on lithium cobalt oxide, providing high energy density and low safety risks when damaged. Lithium ion batteries based on lithium iron phosphate, meanwhile, offer a low energy density, and are safer; making them great in applications such as powering electric tools and medical equipment. Lithium ion batteries offer the best performance to weight ratio, with lithium sulphur batteries providing the highest ratio.

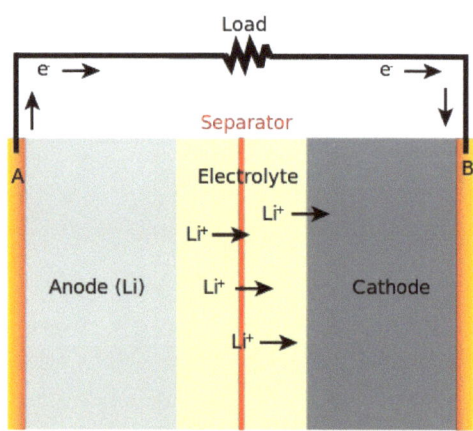

A/B: Current collectors; negative (A), positive (B)

A simple diagram of lithium ion battery (Courtesy Wikimedia Commons)

In lithium phosphate batteries, the cathode is made up of lithium iron phosphate (LiFePO$_4$). Meanwhile, the anode is made up of carbon in the form of graphite. The electrolyte carries positively charged lithium through the separator from the anode to the cathode and vice versa. Movement of lithium ions creates free electrons in the anode, creating a charge at the positive electrode. Then, the electric current flows from the electrode through a powered device (cell phones, computers, etc.) to the negative electrode. The separator blocks the flow of electrons inside the battery.

While the battery is discharging and providing an electric current, the anode generates a flow of electrons from one side to the other. When the battery is charging, however, the opposite happens. The lithium-cobalt oxide in the cathode gives up some of its lithium ions. These ions then move through the electrolyte to the anode and remain there. During this process, the lithium ion battery takes in and stores energy. In both instances, electrons flow in the opposite direction to the ions around the outer circuit.

The movement of ions and electrons are interconnected. If one stops moving, so does the other. If ions stop moving because the battery discharges completely, electrons won't be able to move through the outer circuit. Similarly, if a battery is switched off, the flow of electrons stops and so does the flow of ions. Essentially, the lithium ion battery stops discharging at a high rate; although it does continue discharging at a very slow rate.

Unlike simpler batteries, lithium ion batteries have built-in electronic controllers to regulate how the battery charges and discharges. The electronic controllers prevent lithium ion batteries from overcharging and overheating; thus preventing lithium ion batteries from potentially exploding in some circumstances. The following is some of the attributes of lithium ion batteries:

- **Specific energy:** 100 – 265 W-h/kg
- **Energy density:** 250 - 692 W-h/L
- **Operating temperature:** 5 -45°C
- **Cycle durability:** 400 – 1200 cycles
- **Cost:** $500 to $600 per kWh

Lead Acid Batteries

Lead acid batteries are a cheap and reliable power workhouse used in heavy duty applications. These batteries are typically very large and are always used in non-portable applications such as solar energy storage, vehicle ignition and lights, and backup power because of their colossal weight. Lead-acid batteries are the oldest type of secondary battery and are still very relevant and important in today's world. Despite having very low energy to volume and energy to weight ratios, lead acid batteries have a relatively large power to weight ratio. Consequently, lead acid batteries are able to supply big surge currents when necessary. These attributes, alongside the low costs, make lead acid batteries a wonderful choice to use in several high current applications such as storage in backup power supplies. The properties of lead acid batteries are as follows:

- **Energy Density**: 80-90 Wh/L
- **Specific Power**: 180 W/kg
- **Temperature**: 25°C
- **Cycle durability/life**: 350 cycles
- **Cost**: $300 per kWh

A simple diagram of lead acid battery (Courtesy RoadPro)

* * *

One of the major problems that hinder technological revolutions is power. Battery life affects the success of the deployment of devices. Even though several power management techniques are being adopted to make batteries last longer, a compatible battery must still be selected to achieve such great feats of bringing order to Earth's chaos. As referenced in each type of battery, there are five main factors to be considered to select the correct type of battery:

1. **Energy Density**: Energy density is the total amount of energy that is able to be stored per unit mass or volume. The energy density of a device determines how long the device stays on before it needs a recharge.

2. **Power Density**: Power density is the maximum rate of energy discharge per unit mass or volume.

3. **Safety**: It is important to consider the temperature at which a device will work. At high temperatures, certain battery components will breakdown and can undergo exothermic reactions (reactions accompanied by the release of heat). Generally, high temperatures reduce the performance of most batteries.

4. **Life cycle durability**: The stability of the energy and power density of a battery with repeated cycles are needed for a long battery life, which is required by most applications.

5. **Cost**: It is important that the cost of a battery is correspondent with its performance and will not abnormally increase the overall cost of the project.

Out of these five factors, nickel-metal hydride battery dominates three: energy density, power density, and operating temperature (safety). Thus, nickel-metal hydride would be the most probable candidate to be the chemical configuration used when storing energy. Not only does the nickel-metal hydride battery have the advantage in three factors, but there are several other factors that are of use.

First and foremost, it is environmentally friendly as it has only mild toxins. Not only that, but its nickel content makes recycling profitable. Second, the nickel-metal hydride battery is less prone to the "memory" effect than batteries such as nickel-cadmium batteries, and it can be restored or recharged. Finally, nickel-metal hydride batteries have simple storage and transportation; and are not subject to regulatory control. Their wide operating temperatures also assist in the battery's simplicity in storage and maintenance.

However, nickel-metal hydride batteries do have their flaws. First, these batteries have a low service life as it is reduced by deep discharge. Second, it requires a complex algorithm. Third, nickel-metal hydride batteries are sensitive to overcharge as they do not absorb overcharge well. Thus, trickle charge must be kept low. The fourth flaw is the generation of heat during fast charge and high-load discharge. This is further worsened by the fact that nickel-metal hydride batteries have a high rate of self-discharge. The last flaw in nickel-metal hydride batteries is their low **coulombic efficiency** (the efficiency of the charge by which electrons are transferred) of about 65% compared to the 99% of lithium-ion batteries.

Another problem that arises is the amount of energy that should be stored. Earthquakes are quite unpredictable and inconsistent; and predicting the energy that they would hold is extremely difficult. If the base load of a battery plant is equal to the peak annual amount of energy on the San Andreas fault, then the energy of significant earthquakes of magnitude 8.0 and higher would be thrown away. This, however, is wasteful and thus expensive. Meanwhile, if the base load of a battery plant is equal to the bottom of the curve and use oil and natural gas to compensate for energy use. These, however, are more expensive than the base line and are less environmentally friendly; thus breaking two of the factors that originally would've fit the nickel-metal hydride battery. A third option is to set the base line somewhere in the middle of the curve and use the excess energy during the peak with some loss of efficiency. This third option is the most feasible one to work with, and yet one more question yet arises: how much energy should we store? That is a question that is difficult not because of the different variables that apply in such a large scale, but rather because of the chaotic nature of earthquakes.

In southern California, there are about 10,000 earthquakes a year. Based on the distribution of earthquakes and the average time of an earthquake, a battery plant would require over 42.6 billion kWh. However, with the coulombic efficiency of nickel-metal hydride batteries, the battery plant would store nearly 27.7 billion kWh. This large amount of energy would mean that the battery plant would have to cost between 35.5 and 71.007 million dollars.

Thus, it is possible, even with our current technology, to bring order to Earth's chaos. Despite the many flaws that could arrive with our technology, we must ask our last question, which we shall depend on future technologies: can we prolong the effects of earthquakes?

This question, while seemingly absurd and unreasonable, assists in storing energy as it helps lower the energy density and thus, the cost. While further research is required as we discover new ways to generate energy, there is one way to utilize the conservation of momentum: Newton's Cradle. While our current technology falls short in technology that utilizes the conservation of momentum, our current studies do indeed point to the direction of technologies similar to Newton's Cradle. This is only one way in which we can depend on bringing order to Earth's chaos. This is only one way in which we can live in a greener future.

Conclusion

THERE ARE MANY OBSTACLES when bringing order to Earth's chaos, the most prominent of which being the limitation of our own technology. This fascinating journey explains one way in which we can take opportunities nature has given us. There is much to debate about the legitimacy and practicality of such a seemingly impossible feat, but this opportunity is one that is almost impossible to pass. This opportunity is one that could open the door to independence from fossil fuels.

The exploitation of fossil fuels infamously has a heavy toll on ecosystems. As of 2018, we have caused animal species to go extinct at a rate 1,000 times faster than the natural rate of extinction, with 60% of animal populations disappearing since 1970. While this rate is not caused entirely by fossil fuel, it is interconnected with an equally terrifying and depressing consequence: in 2014, fossil fuel emissions account for 91% of global CO_2 emissions from human sources. These emissions undoubtedly have deadly consequences for humans, for nature, and for the planet we live in. In fact, 12.6 million people in the US alone are exposed to toxic air pollution every day. We have exploited so much fossil fuel that air pollution alone caused more deaths than smoking. The question is: what are we going to do about it?

We are already making amazing progress to move to a greener future. We have started to exploit the water, the wind, and the Sun in green and renewable fashion that undoubtedly caused a bit more independence from non-renewable resources. Despite this, we must be aware. We must be aware not because of the inconsistency of the renewable resources, but because of the drawbacks. Our technology for energy storage is indeed a major challenge for many renewable resources, including the resource which I have proposed: earthquakes. However, even with the required technology in place, we would require more energy to depend on renewable energy.

Over 50% of people currently live in urban areas, with the percentage only increasing in the future. Unfortunately, many of these people do not have access to renewable energy. Perhaps they live in regions with little wind, live near the poles where there is relatively little sunlight, or simply live in apartments where they do not have access to a roof to place solar panels. This is where the utilization of earthquakes comes in. While there are many places in which there is little to no earthquakes, such as in the Arabian Peninsula, most areas of the world, including nearly all the coasts in the world, frequently experience earthquakes.

With coasts being hotspots of earthquakes, let us not forget that most people live near the coast; meaning that earthquakes not only have one of the highest energy potential of any renewable resource, but also the most consistency. For the great challenges we shall face to bring order to Earth's chaos, let us aim to utilize the energy and consistency of earthquakes. Fossil fuels will eventually deplete, but the Earth's rage and chaos will persist. There is much to be worked on, so let that be the focus of our future: not to bring chaos from Earth's order, but to bring order from Earth's chaos.

Glossary

A

Amplitude: The maximum difference of the potential from the average value.

Anode: the terminal in which an electric current (see **electric current**) flows in from outside.

Archipelago: A vast group of islands.

Armorica: A microcontinent (see **microcontinent**) that rifted away from Gondwana (see **Gondwana**) towards the end of the Silurian period (443 to 416 million years ago) and collided with Laurussia (see **Laurussia**) towards the end of the Carboniferous period (359 to 299 million years ago).

Asthenosphere: The upper layer of the earth's mantle where there is relatively low resistance to plastic flow and convection is believed to occur.

Avalonia: A microcontinent (see **microcontinent**) during the Paleozoic Era (541 to 245 million years ago) that collided with Gondwana (see **Gondwana**) to from the interior of Pangaea (see **Pangaea**).

B

Back-arc basin: A rock formation where a series of layers of rock in the ground dip into the center, formed by a process called back-arc spreading.

Back-arc spreading: A process that begins when one tectonic plate descends under another.

Baltica: A paleocontinent (see **paleocontinent**) that was formed in the Paleoproterozoic (2.5 to 1.6 billion years ago), with its thick core being over three billion years old, forming as a part of the Rodinia (see **Rodinia**) supercontinent.

Baryosphere: The central core of the Earth.

Basalt: A dark, fine-grained volcanic rock formed from the quick cooling of lava rich in magnesium and iron that is exposed at or very near to the surface.

Battery: A self-contained device that can convert a limited amount of chemical energy into electricity.

C

Carboniferous: Carbon-bearing.

Cathode: The terminal in which electric current (see **electric current**) flows out.

Central Pangean Mountains: A vast mountain range in the middle of Pangaea during the Triassic period (251 to 199 million years ago).

Chaos System: The system that includes the external processes of the Earth, such as earthquakes, volcanic eruptions, or the movement of tectonic plates.

Chemical potential: The ability of an uncharged atom or molecule to work

Cimmeria: An ancient continent that drifted apart from Gondwana in the southern hemisphere and was accreted to Eurasia in the northern hemisphere.

Compressional waves: Waves that have alternating pulses like sound waves.

Continental craton: An old and stable part of the continental lithosphere (see **lithosphere**).

Convection currents: Currents in a fluid caused by the tendency of hotter materials to move upwards while colder materials sink under below.

Convergent boundary: A boundary in which two tectonic plates collide.

D

Deccan traps: A large igneous province (see **igneous**) in the Deccan Plateau in west-central India.

Divergent boundary: A boundary in which two tectonic plates move away from each other.

E

Electric current: The rate in which electric charge flows past a point.

Electric discharge: The release and transmission of electricity through a channel.

Electrodes: Conductors through which electricity enters or leaves an object.

Electrolytes: A chemical channel that allows the flow of electric charge from the cathode to the anode.

Epicenter: The point on top of the focus.

Epicontinental: Areas of sea on top of continental shelves.

Estuary: The place in which the tide meets the stream.

F

Fault: A fractured weak point that marks the contact and rupture of two plates.

Flood basalts: The result of huge volcanic eruptions that cover a vast area with volcanic basalts (see **basalt**).

Focus: The point in which earthquakes form.

G

Gondwana: A supercontinent (see supercontinent) that existed from the Neoproterozoic to the Jurassic (550 to 180 million years ago). There are many debates as to how Gondwana split up has been up to debate, but recent research shows that it may have been split in two.

I

Iapetus Ocean: An ocean in the southern hemisphere between Laurentia, Baltica, and Avalonia (see **Laurentia – Baltica – Avalonia**) that existed from the late Neoproterozoic era to the early Paleozoic era (600 to 400 million years ago).

Igneous: Rocks solidified from lava or magma.

Indochina: A peninsula in southeastern Asia, made up of Vietnam, Cambodia, Laos, Thailand, western Malaysia, and Myanmar.

Interglacial: A geological gap of warmer temperature than the average global temperature lasting for a few thousand years.

Inverse: Opposite in direction, position, or order.

Inverse-square law: A physical law that states that a physical quantity is inversely proportional (see **inverse**) to the square of the distance from the source.

Ion: An atom with an electric charge due to gaining or losing electrons.

Iridium: A very hard, brittle, silvery-white metal rare on Earth but common in meteorites.

Isostasy: The state of gravitational equilibrium between the crust and the mantle so that the crust "floats" on top of the mantle.

K

Kinetic friction force: A force that acts between two or more moving surface.

L

Laurasia: A supercontinent (see **supercontinent**) in the Northern Hemisphere around 200 million years ago that consists of North America, Greenland, Europe, and Asia with the exception of the Indian subcontinent.

Laurentia: A large continental craton (see **continental craton**) that forms the geological core of the North American continent.

Laurussia: A minor supercontinent which formed in the Devonian period (419.2 to 358.9 million years ago), consisting of much of northern Europe, Greenland, and North America.

Limestone: A hard sedimentary rock (see **sedimentary**) used as building material and in making cement.

Lithosphere: The outer part of the Earth, which is made up of the crust and the upper part of the mantle.

Love wave: Waves that shake violently from side to side

Low-speed layer: A layer in which the speed of seismic waves drops significantly.

M

Major plates: Tectonic plates with an area larger than 20 million sq. kilometers.

Mantle convection: The process in which the Earth's mantle moves due to convection currents (see **convection currents**) carrying heat from the Earth's interior to the surface.

Microcontinent: An isolated piece of continental crust.

Microplates: Tectonic plates with an area smaller than 1 million sq. kilometers.

Mid-oceanic ridge: Divergent boundaries (see **divergent boundary**) between two oceanic plates.

Minor plates: Tectonic plates with an area between 1 and 20 million sq. kilometers.

Mohorovicic discontinuity: The sharp change to denser rocks between the crust and the mantle.

Moment magnitude scale: A type of magnitude scale used by seismologists to measure the size of earthquakes compared to the energy they release.

O

Order System: The system that consists of the internal processes of the Earth, such as mantle convection.

Oxidation reaction: A reaction in which oxygen is gained.

P

Paleocontinent: An area of continental crust that was once a major landmass in the geological past.

Paleo-Tethys Ocean: An ocean located along the northern border of Gondwana (see **Gondwana**) that started to open during the Middle Cambrian and closed during the Late Triassic (504.5 to 199 million years ago).

Pangaea: A supercontinent (see **supercontinent**) that assembled in the late Paleozoic era 335 million years ago and began to break apart in the early Mesozoic era about 175 million years ago.

Pannotia: A relatively short-lived supercontinent that formed at the end of the Precambrian and broke about ~560 million years ago as the Iapetus Ocean (see **Iapetus Ocean**) opened.

Panthalassic Ocean: A super-ocean (see **super-ocean**) that surrounded Pangaea (see **Pangaea**). ~250 million years ago, it occupied almost 70% of Earth's surface.

Peridotite: A dense, coarse-grained rock.

Plate boundary: The boundary area between two or more tectonic plates.

Polydimethylsiloxane: The simplest member of the group of silicone polymers.

Polyester: A type of large molecules composed of many repeated subunits that contain the ester functional group in their main chain.

Proto-Atlantic Ocean: See **Iapetus Ocean**.

Pyrosphere: The middle part of the Earth, also known as the mantle.

R

Rayleigh wave: A wave that rolls like the waves of the ocean.

Rheic Ocean: An ancient ocean which separated two paleocontinents, Gondwana and Laurussia (see **Gondwana – Laurussia**).

Rhine River: One of the major European rivers which has its sources in Switzerland and flows in a mostly northerly direction, emptying into the North Sea.

Richter scale: A scale of numbers used to tell the size of earthquakes.

Rodinia: A supercontinent (see **supercontinent**) in the Neoproterozoic era that had most, if not all of Earth's landmass, existing from 1.1 billion to 750 million years ago.

S

Seabed: The ocean floor.

Sedimentary: A rock formed from sediment deposited by water or air.

Sedimentary deposits: Mineral deposits formed during the build-up of sediments on the seabed (see **seabed**).

Seismic waves: Vibrations triggered by earthquakes.

Shearing waves: An alternate side-to-side wobble travelling across a body, like jelly.

Silica: Any compound made of a chemical combination of silicon and oxygen.

Silicates: Rocky materials mainly comprised of silicon and oxygen.

Static friction: The force necessary to keep objects at rest.

Sulphur dioxide: A toxic gas with a foul, sharp smell that reacts easily with other substances to form harmful compounds, such as sulfuric acid.

Supercontinent: The assembly of most or all of Earth's landmass to form a single massive landmass.

Super-ocean: An ocean that surrounds a supercontinent (see **supercontinent**).

Surface wave: A wave that travels along the surface of the Earth.

T

Tectonic plates: A description of how the Earth's crust is divided into continental-sized plates jostle around; creating volcanoes and mountain chains.

Tethys Ocean: An ocean between the supercontinents (see **supercontinent**) of Gondwana and Laurasia (see **Gondwana – Laurasia**) during most of the Mesozoic Era before the opening of the Indian and Atlantic oceans during the Cretaceous Period (145 to 66 million years ago).

Thames River: A river that flows through southern England.

Transform boundary: A boundary in which plates slide passed each other.

Triboelectric effect: An effect in which certain materials become electrically charged after coming into contact with another different material and are then separated.

Triboelectric generator: Generators based on triboelectric effect (see **triboelectric effect**).

Tsunami: A large sea wave caused by an earthquake or other disturbances underwater.

Y

Yaodong: A kind of cave dwelling in Loess Plateau in northeastern China.

Index

A

Amplitude 50, 61, 63, 66
Antioch..27
Armorica 11
Asthenosphere........................... 44
Atmosphere 43
Avalonia 11, 14

B

Baltica 10, 12, 14, 16, 17
Baryosphere......................... 43, 44
Battery... 64

C

Cimmeria 20
Compressional waves 39
Continental crust 38, 39
Convergent boundaries....... 47, 57
Core 38, 41, 42, 44
Crust 38, 39, 40, 43, 45, 46, 47, 48, 49

D

Deccan Traps 23
Divergent boundaries 48, 57

E

Earthquake.....6, 27, 28, 29, 30, 31, 32, 33, 34, 47, 50, 51, 52, 53, 54, 56, 58, 59, 60, 62
Epicenter............ 28, 32, 33, 50, 51
European Macroseismic scale ...52

F

Fault 30, 49, 50, 53, 58, 62
Fault lines 49, 58

Focus.......46, 50, 51, 54, 56, 60, 61

G

Glaciation 18, 25
Gondwana .10, 11, 12, 14, 16, 17, 19
Great Shake 30, 58, 60, 62

I

Iapetus Ocean............ 10, 12, 14, 16
Iberia.................................... 11, 16
Inner core...........................42, 44
Inverse-square law 61
Isostasy...................................... 39

K

Kinetic friction...........................59

L

Laurasia....................................... 22
Laurentia 10, 12, 14, 16, 17, 19
Lithosphere 43, 44, 45
Love waves 50
Low-speed layer.......................44

M

Major plates 37, 45
Mantle 38, 39, 40, 41, 43, 44
Mantle convection 37, 44
Microplates 45
Mid-oceanic ridges 49
Minor plates................................45
Modified Mercalli scale..............51
Mohorovicic discontinuity.. 37, 40
Moment magnitude scale 32, 53, 60

O

111 | P a g e

Oceanic crust 39
Outer core 42, 44

P

Paleo-Tethys Ocean .. 12, 14, 17, 20
Pangaea ... 17, 18, 19, 20, 21, 22, 23
Pannotia 10
Panthalassic Ocean ... 10, 12, 14, 18
Peridotite 41
Proto-Atlantic Ocean 22
Pyrosphere 43, 44

R

Rayleigh waves 50
Rheic Ocean 12, 14, 17
Richter scale 26, 28, 31, 47, 50, 51, 53

S

San Andreas fault 49, 58
Seismic waves ... 39, 40, 44, 50, 61
Shearing waves 39, 42
Silica 22, 41, 43, 44
Silicates 41, 42
Static friction 58, 59
Surface waves 50, 51

T

Tectonic plates 26, 27, 37, 45, 46, 47, 56, 57, 58, 66
Tectonic plates 37, 45, 47
Tethys Ocean 17, 18, 20, 22, 23
Transform boundaries . 49, 58, 66
Triboelectric generators 63
Tsunami 29, 33, 34, 54

Y

Yaodongs 28

Bibliography

Bates, Mary. How does a battery work? - MIT School of Engineering. 1 May 2012.

Coontz, Robert. ScienceMag. March 15 2011.

Denker, John. How to Define Anode and Cathode. 2004.

Environmental Protection Agency. 3 April 2018. <https://www.epa.gov/energy/electricity-storage>.

Northwestern Edu. Power System. n.d.

Rathi, Akshat. The world's largest "virtual battery plant" is now operating in the Arabian desert - Quartz. 30 January 2019.

USGS. Earthquake Magnitude, Energy Release, and Shaking Intensity. n.d.

Energystorage.org. (2019). Sodium Sulfur (NaS) Batteries | Energy Storage Association.

Woodford, C. (2018). How do lithium-ion batteries work?. [online]

Energy.gov. (2019). How Does a Lithium-ion Battery Work?. [online]

Berry, S. (2017, January 13). Tag: Cycle Life.

LITHIUM BATTERY ADVANTAGES. (n.d.). Retrieved from https://relionbattery.com/technology/why-choose-lithiumLiaoning Borui Machinery Co. Cast iron price calculator. Retrieved from http://www.iron-foundry.com/cast-iron-price-calculator.html

Song Y., Cheng X. L., Han M., Meng B. (2016, June) A flexible large-area triboelectric generator by low-cost roll-to-roll process for location-based monitoring.

Matousek, M. (2019, 31 July) Tesla Just Announced a Giant New Battery System to Store Renewable Energy. Retrieved from https://www.sciencealert.com/tesla-just-announced-a-giant-new-battery

IRIS, (2011). How Often Do Earthquakes Occur? Education and Outreach Series(3).

Wong, W.G. (2014, 31 January) Harvesting Power Using Triboelectric Generators.

Backhouse, Frid, et al. "Natural." *501 Most Devastating Disasters - Updated Edition*, edited by Emma Hill, Polly Manguel, Bounty Books, 2011, pp. 8–88.

Rother, David A. *Geology: A Complete Introduction.* Hodder Education, 1997

About the Author

Alwaleed Alghanim is an author born in January 27th, 2002 who studies science for a living. Born in the Eastern Province of Saudi Arabia, Alwaleed Alghanim grew up in a family of five, with two supportive parents, an older brother, and a younger sister. With a passion for writing, Alwaleed aspires to show the future generations the beauty of science.